Managing Editor
Karen Goldfluss, M.S. Ed.

Editor-in-Chief
Sharon Coan, M.S. Ed.

Cover Artist
Barb Lorseyedi

Art Coordinator
Kevin Barnes

Art Director
CJae Froshay

Imaging
James Edward Grace
Rosa C. See

Product Manager
Phil Garcia

Decimals & Money

GRADES 3 & 4

Publishers
Rachelle Cracchiolo, M.S. Ed.
Mary Dupuy Smith, M.S. Ed.

Authors

Teacher Created Materials Staff

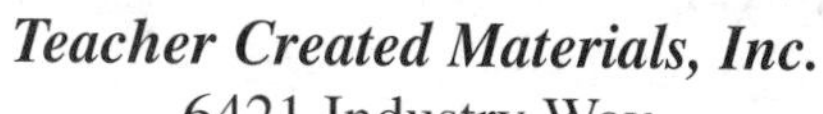

Teacher Created Materials, Inc.
6421 Industry Wav
Westminster, CA 92
www.teachercreated
ISBN-0-7439-332

Table of Contents

Introduction

The old adage "practice makes perfect" can really hold true for your child and his or her education. The more practice and exposure your child has with concepts being taught in school, the more success he or she is likely to find. For many parents, knowing how to help their children can be frustrating because the resources may not be readily available. As a parent it is also difficult to know where to focus your efforts so that the extra practice your child receives at home supports what he or she is learning in school.

This book has been designed to help parents and teachers reinforce basic skills with their children. *Practice Makes Perfect* reviews basic math skills for children in grades 3 and 4. The math focus is on decimals and money. While it would be impossible to include all concepts taught in grades 3 and 4 in this book, the following basic objectives are reinforced through practice exercises. These objectives support math standards established on a district, state, or national level. (Refer to the Table of Contents for the specific objectives of each practice page.)

- using models to identify decimals
- identifying place value in decimals
- reading decimals
- comparing and ordering decimals
- adding decimals
- subtracting decimals
- rounding decimals
- estimating decimal sums
- counting, adding, subtracting, and multiplying money
- estimating sums and differences of money

There are 36 practice pages organized sequentially, so children can build their knowledge from more basic skills to higher-level math skills. (**Note:** Have children show all work where computation is necessary to solve a problem. For multiple choice responses on practice pages, children can fill in the letter choice or circle the answer.) Following the practice pages are six practice tests. These provide children with multiple-choice test items to help prepare them for standardized tests administered in schools. As your child completes each test, he or she should fill in the correct bubbles on the answer sheet (page 46). To correct the test pages and the practice pages in this book, use the answer key provided on pages 47 and 48.

How to Make the Most of This Book

Here are some useful ideas for optimizing the practice pages in this book:

- Set aside a specific place in your home to work on the practice pages. Keep it neat and tidy with materials on hand.
- Set up a certain time of day to work on the practice pages. This will establish consistency. An alternative is to look for times in your day or week that are less hectic and conducive to practicing skills.
- Keep all practice sessions with your child positive and constructive. If the mood becomes tense, or you and your child are frustrated, set the book aside and look for another time to practice with your child.
- Help with instructions if necessary. If your child is having difficulty understanding what to do or how to get started, work the first problem through with him or her.
- Review the work your child has done. This serves as reinforcement and provides further practice.
- Allow your child to use whatever writing instruments he or she prefers. For example, colored pencils can add variety and pleasure to drill work.
- Pay attention to the areas in which your child has the most difficulty. Provide extra guidance and exercises in those areas. Allowing children to use drawings and manipulatives, such as coins, tiles, game markers, or flash cards, can help them grasp difficult concepts more easily.
- Look for ways to make real-life applications to the skills being reinforced.

Practice 1

1. What decimal matches the *shaded* part of this rectangle?

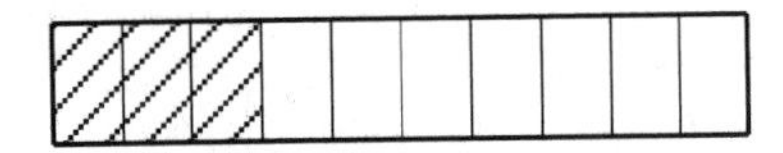

(A) 0.3 (B) 3.7 (C) 3 (D) 0.03

2. What decimal matches the *shaded* part of these rectangles?

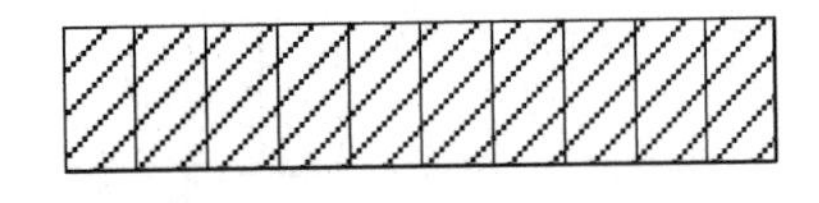

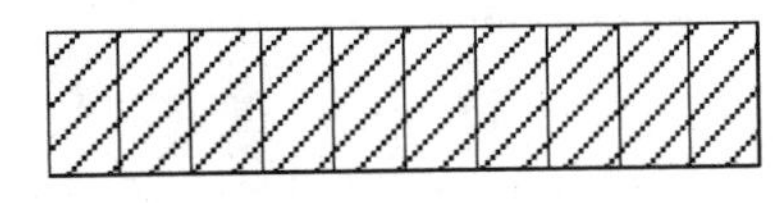

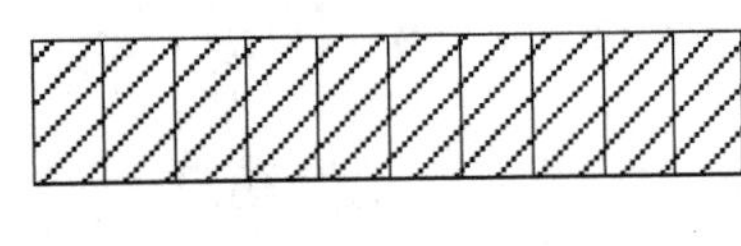

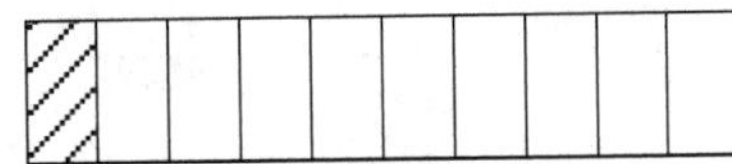

(A) 31.0 (B) 3.1
(C) 0.31 (D) 3.01

3. What decimal matches the *shaded* part of this rectangle?

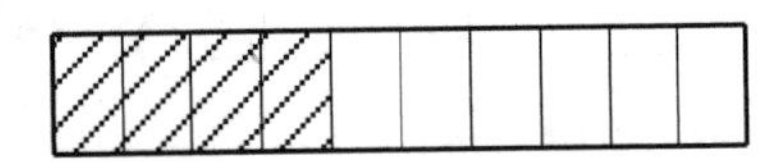

(A) 4 (B) 0.04 (C) 0.4 (D) 4.6

4. What decimal matches the *shaded* part of this rectangle?

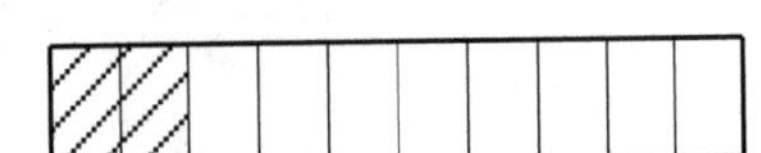

(A) 2.8 (B) 2 (C) 0.02 (D) 0.2

5. What decimal matches the *shaded* part of this rectangle?

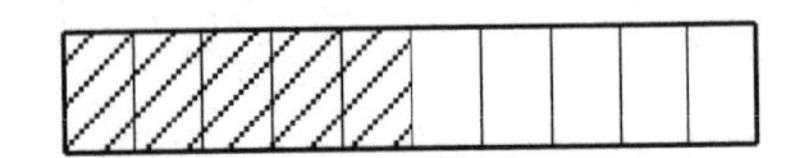

(A) 5 (B) 0.05 (C) 0.5 (D) 5.5

6. What decimal matches the *shaded* part of these rectangles?

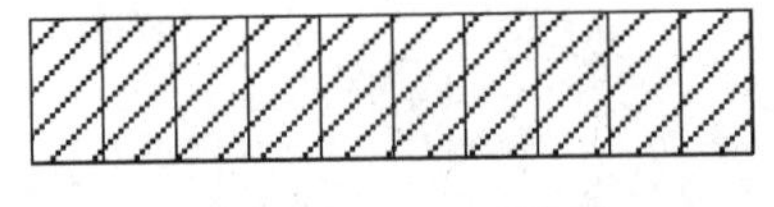

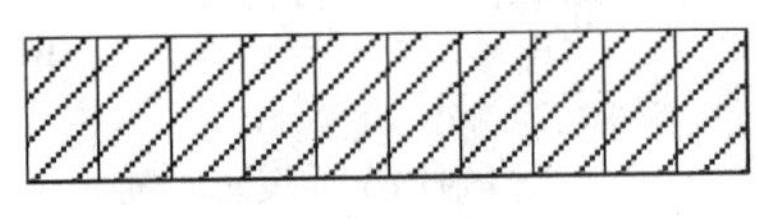

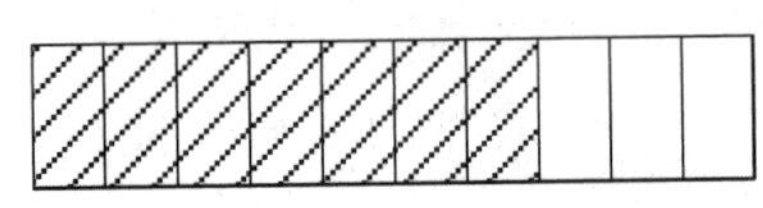

(A) 2.07 (B) 27.0
(C) 2.7 (D) 0.27

7. What decimal matches the *shaded* part of this rectangle?

(A) 0.07 (B) 0.7 (C) 7 (D) 7.3

8. What decimal matches the *shaded* part of this rectangle?

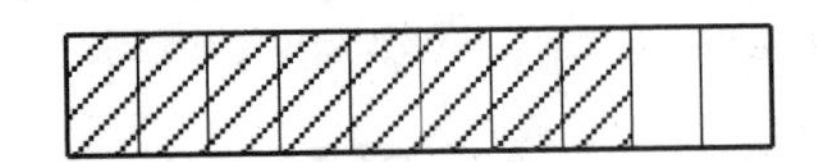

(A) 0.8 (B) 8 (C) 8.2 (D) 0.08

9. What decimal matches the *shaded* part of this rectangle?

(A) 1.9 (B) 0.01 (C) 1 (D) 0.1

10. What decimal matches the *shaded* part of this rectangle?

(A) 9 (B) 9.1 (C) 0.09 (D) 0.9

Practice 2

1. What decimal matches the *shaded* part of these rectangles?

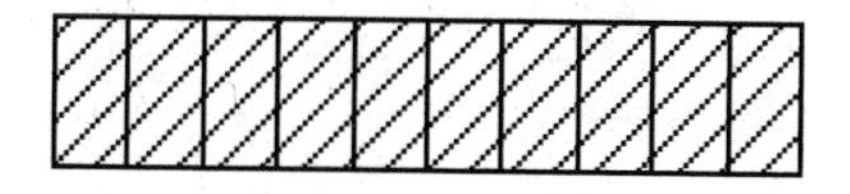

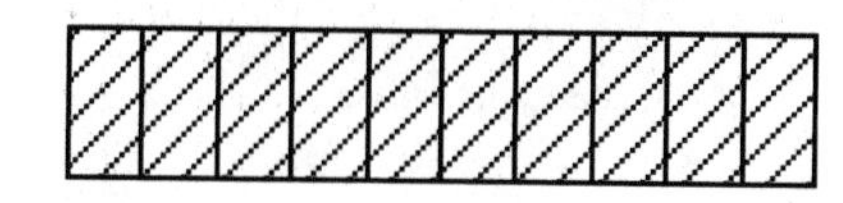

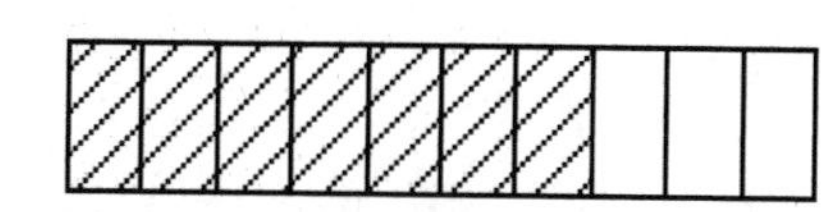

(A) 0.37 (B) 3.7
(C) 37.0 (D) 3.07

2. What decimal matches the *shaded* part of these rectangles?

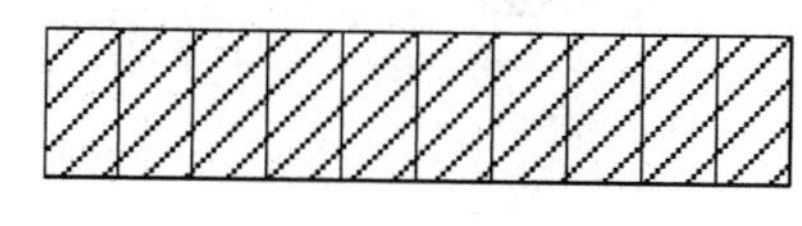

(A) 17.0 (B) 1.07
(C) 1.7 (D) 0.17

3. What decimal matches the *shaded* part of these rectangles?

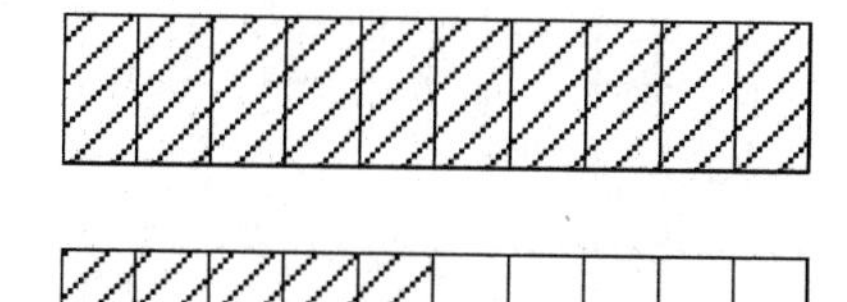

(A) 1.5 (B) 15.0
(C) 1.05 (D) 0.15

4. What decimal matches the *shaded* part of these rectangles?

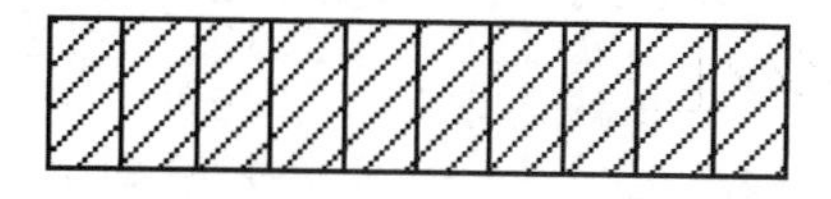

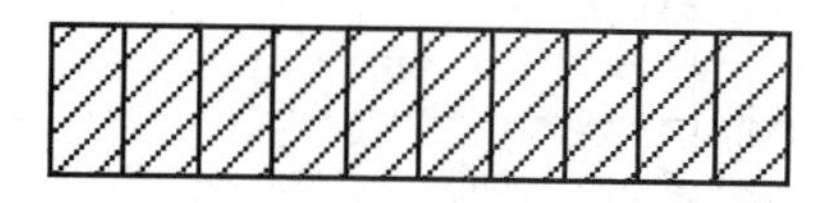

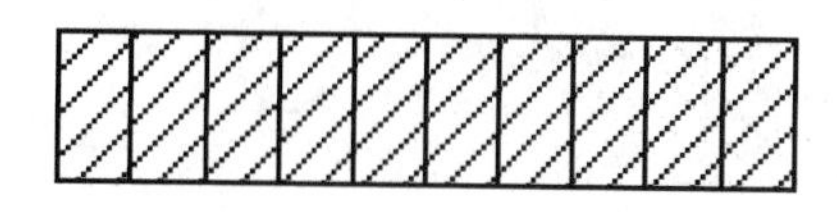

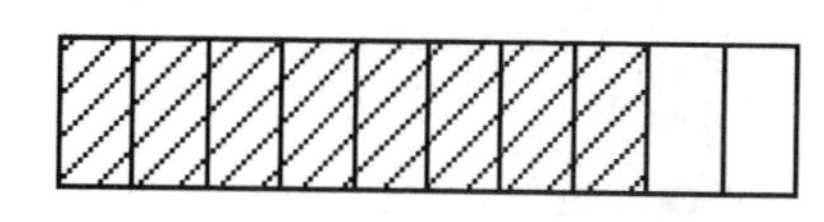

(A) 0.38 (B) 3.8
(C) 3.08 (D) 38.0

5. What decimal matches the *shaded* part of these rectangles?

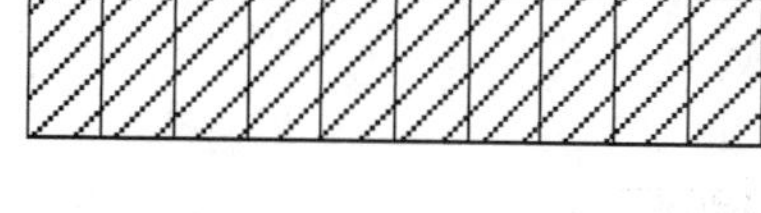

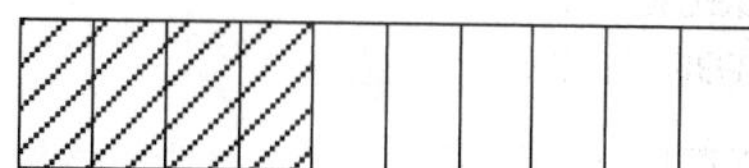

(A) 14.0 (B) 0.14
(C) 1.04 (D) 1.4

6. What decimal matches the *shaded* part of these rectangles?

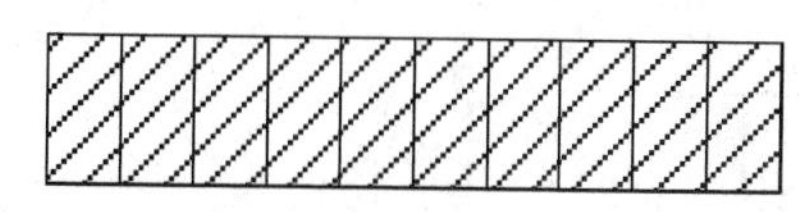

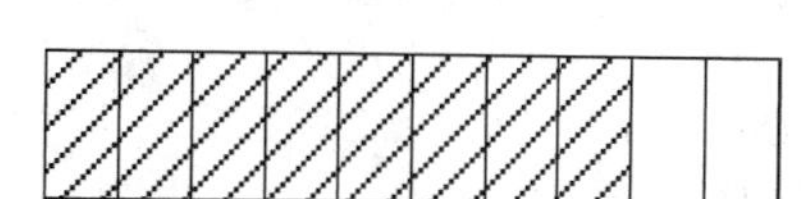

(A) 1.08 (B) 1.8
(C) 0.18 (D) 18.0

Practice 3

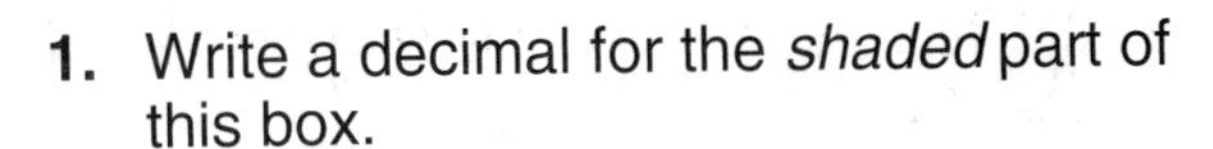

1. Write a decimal for the *shaded* part of this box.

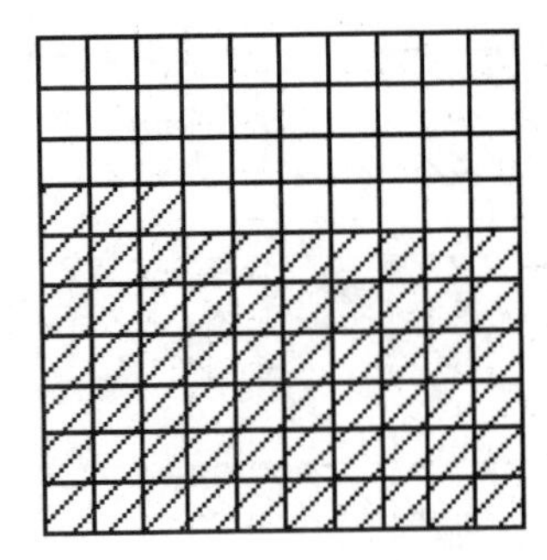

____

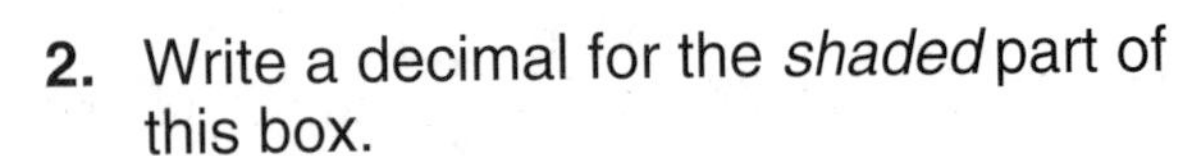

2. Write a decimal for the *shaded* part of this box.

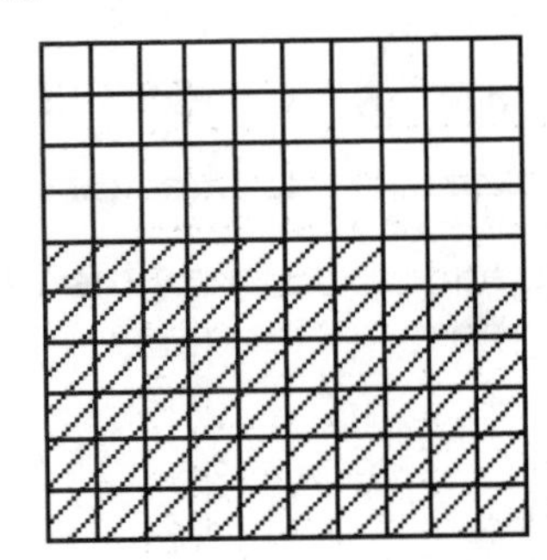

____

3. Write a decimal for the *shaded* part of this box.

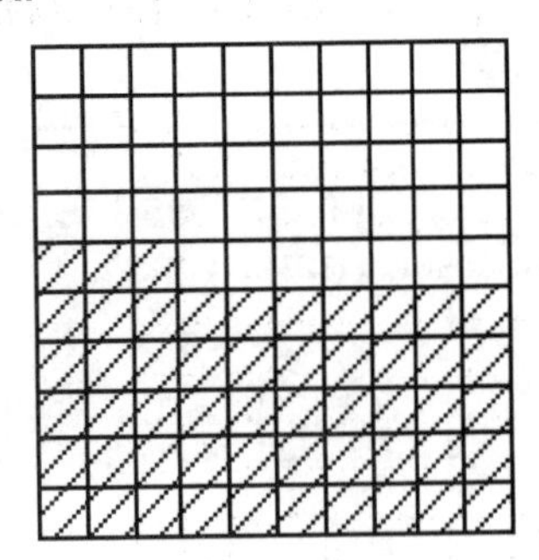

____

4. Write a decimal for the *shaded* part of this box.

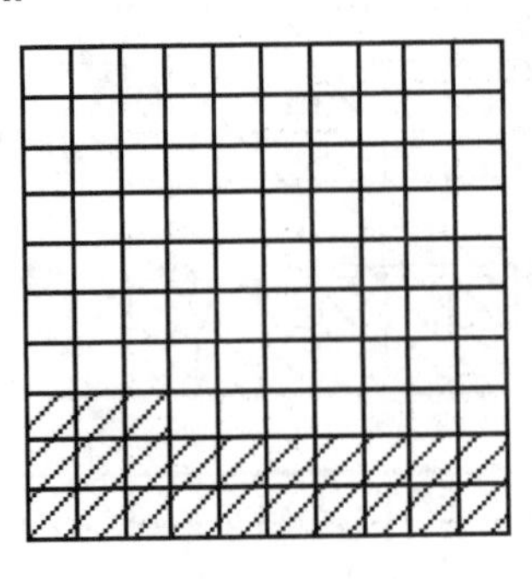

____

5. Write a decimal for the *shaded* part of this box.

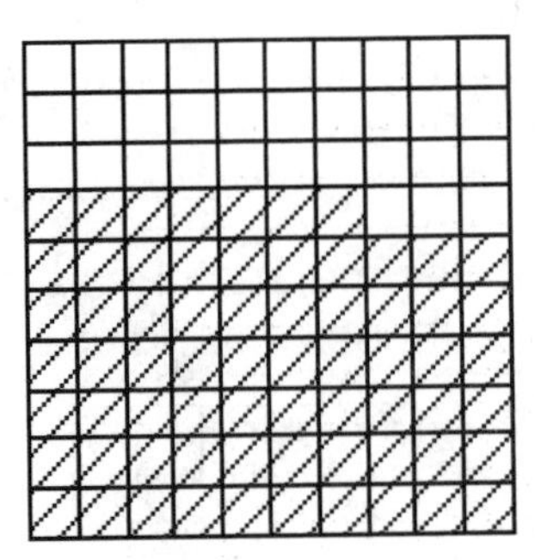

____

6. Write a decimal for the *shaded* part of this box.

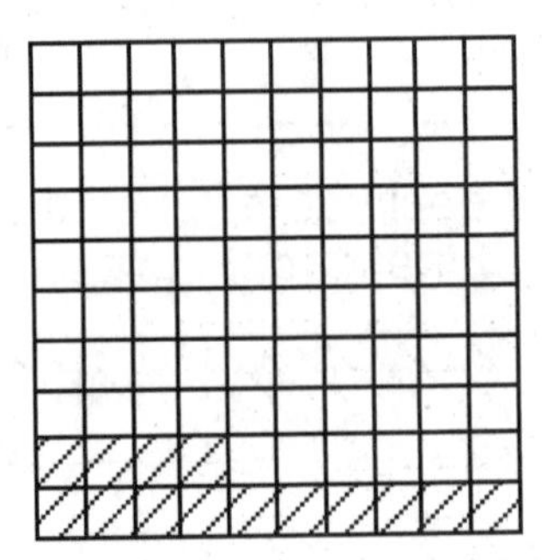

____

Practice 4

1. Write a decimal for the *shaded* part of this box.

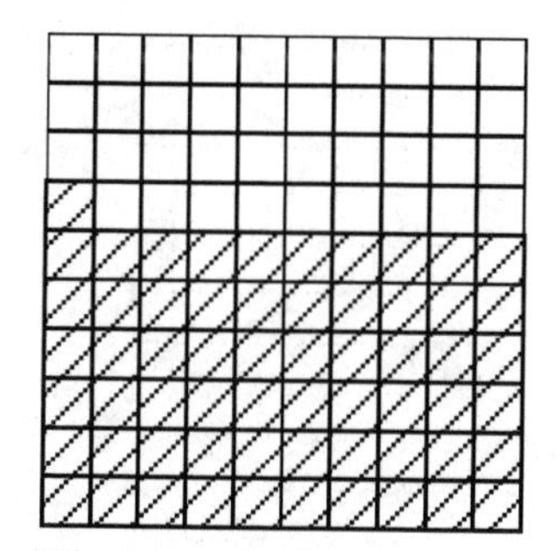

(A) 0.28 (B) 0.72
(C) 0.61 (D) 0.39

2. Write a decimal for the *shaded* part of this box.

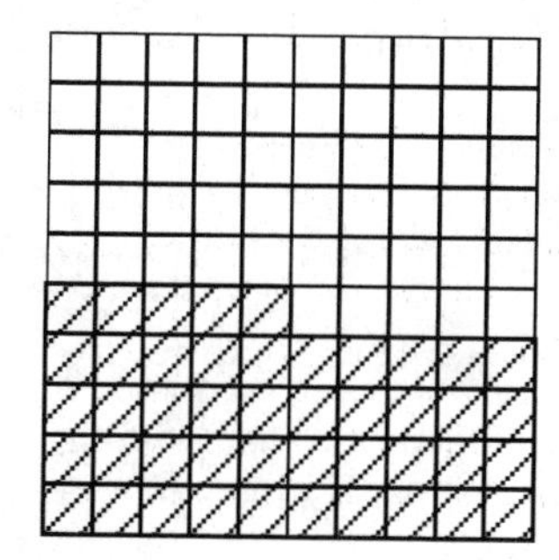

(A) 0.55 (B) 0.45
(C) 0.35 (D) 0.65

3. Write a decimal for the *shaded* part of this box.

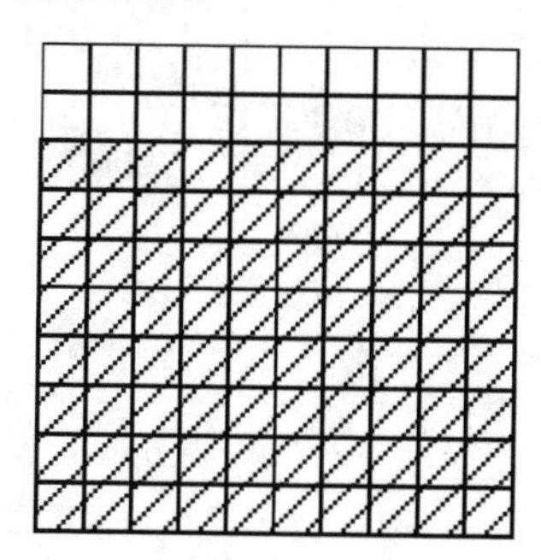

(A) 0.89 (B) 0.11
(C) 0.79 (D) 0.21

4. Write a decimal for the *shaded* part of this box.

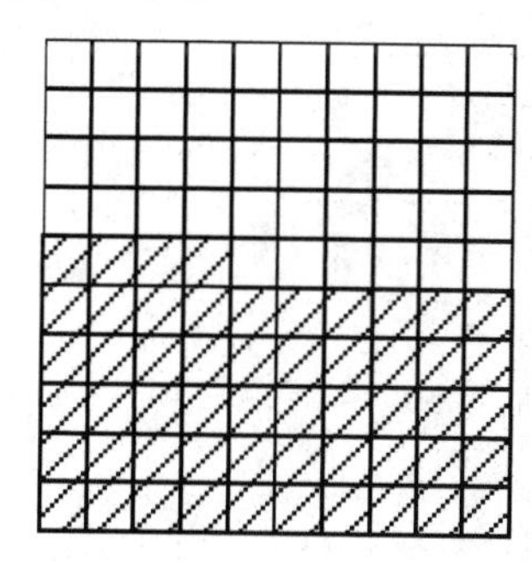

(A) 0.54 (B) 0.55
(C) 0.45 (D) 0.46

5. Write a decimal for the *shaded* part of this box.

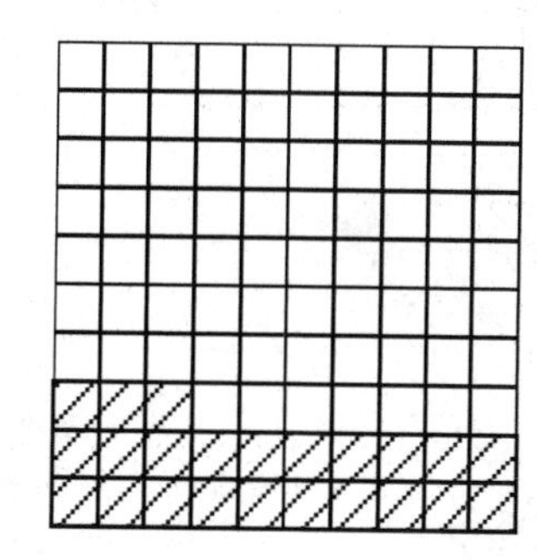

(A) 0.22 (B) 0.78
(C) 0.77 (D) 0.23

6. Write a decimal for the *shaded* part of this box.

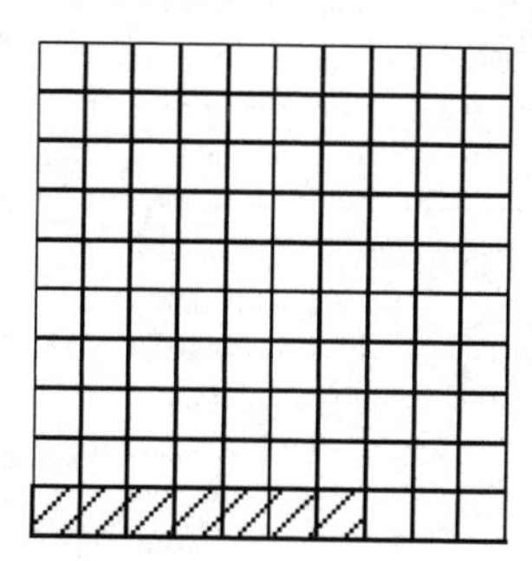

(A) 0.82 (B) 0.93
(C) 0.18 (D) 0.07

write as a fraction

Practice 5

1. What is the decimal number for the *shaded* portion of the circle?

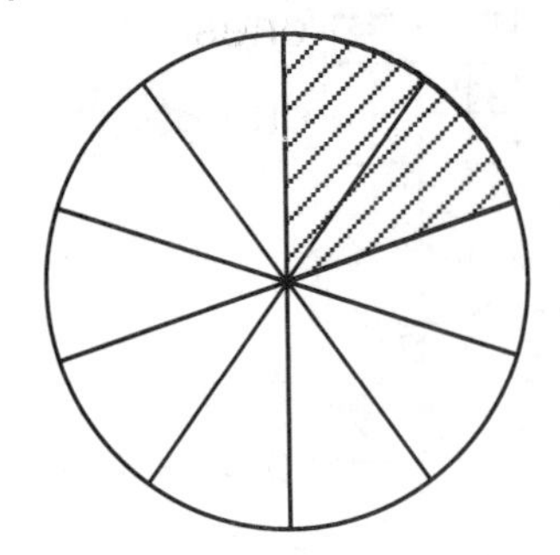

____

2. What is the decimal number for the *shaded* portion of the circle?

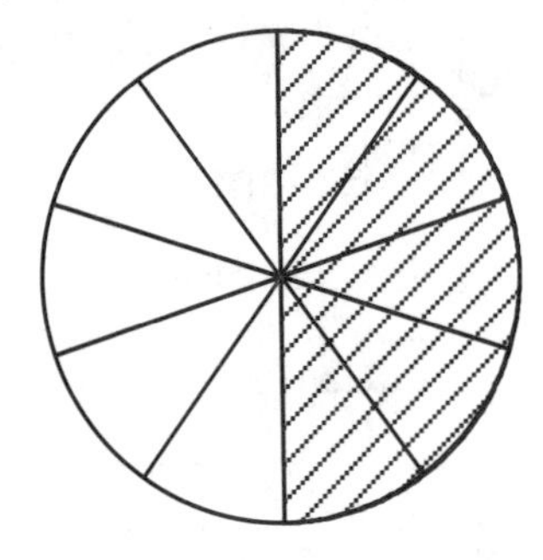

____

3. What is the decimal number for the *shaded* portion of the circle?

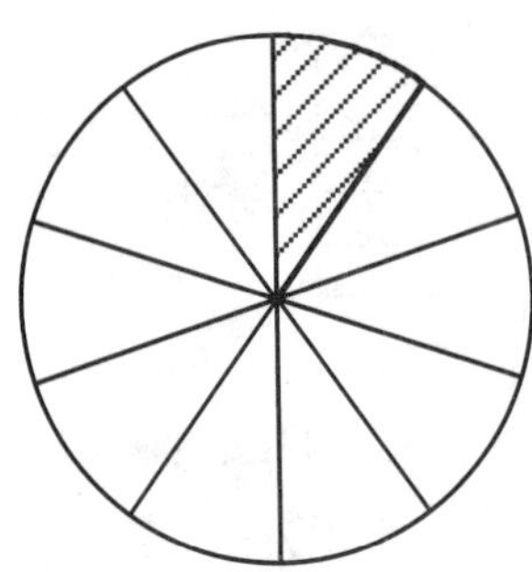

____

4. What is the decimal number for the *shaded* portion of the circle?

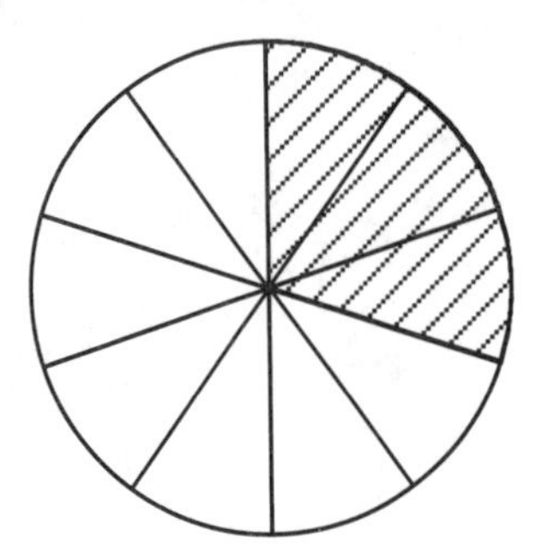

____

5. What is the decimal number for the *shaded* portion of the circle?

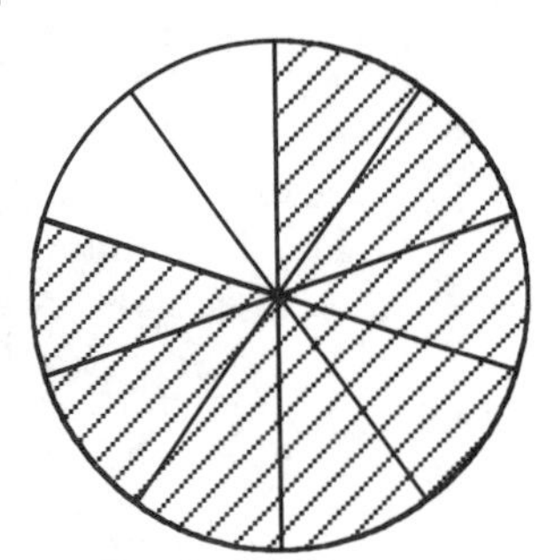

____

6. What is the decimal number for the *shaded* portion of the circle?

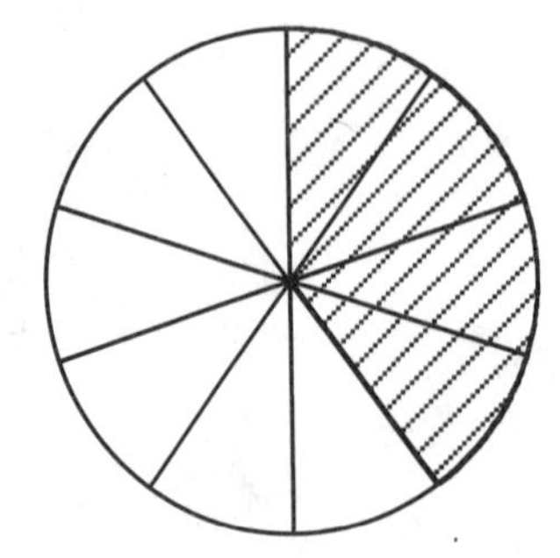

____

7. What is the decimal number for the *shaded* portion of the circle?

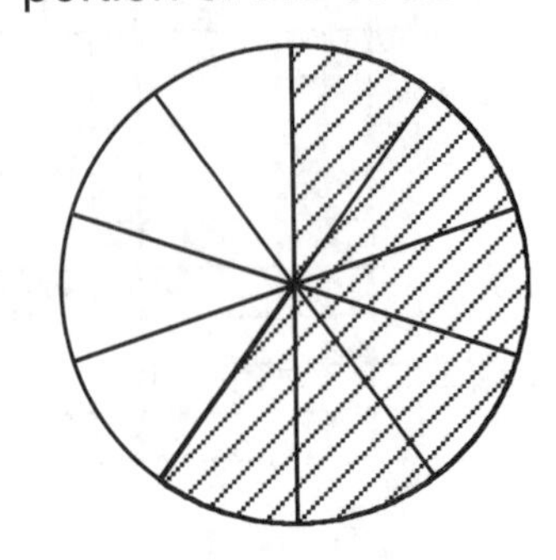

____

8. What is the decimal number for the *shaded* portion of the circle?

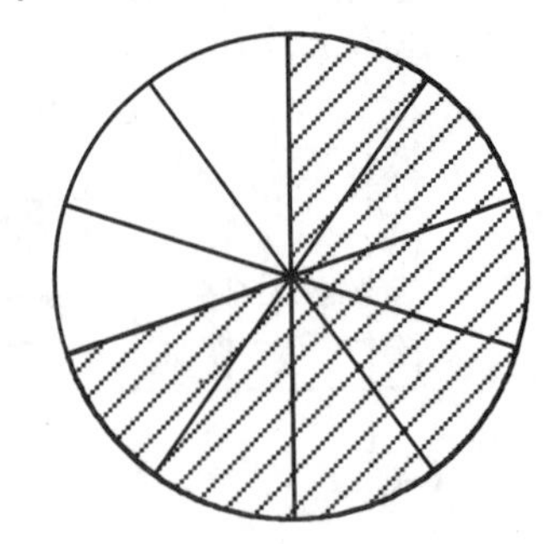

____

9. What is the decimal number for the *shaded* portion of the circle?

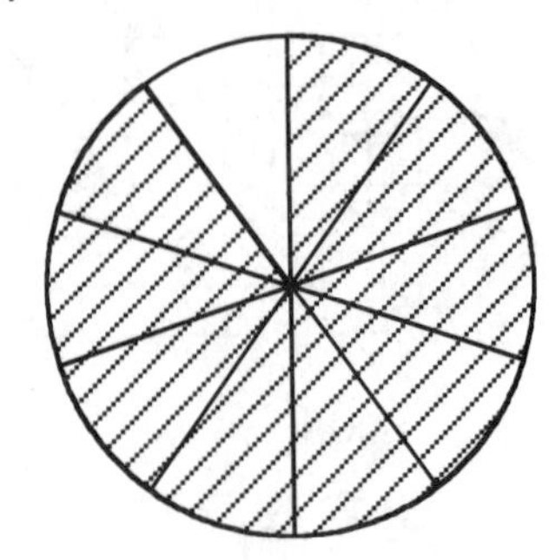

____

Practice 6

1. What is the place value of the 8 in 857.01?
   (A) tenths (B) hundreds
   (C) tens (D) hundredths

2. What is the place value of the 9 in 432.96?
   (A) hundredths (B) tenths
   (C) hundreds (D) tens

3. What is the place value of the 4 in 942.85?
   (A) tens (B) ones
   (C) hundreds (D) tenths

4. What is the place value of the 3 in 107.63?
   (A) tens (B) hundreds
   (C) hundredths (D) tenths

5. What is the place value of the 9 in 617.94?
   (A) hundreds (B) tens
   (C) tenths (D) hundredths

6. What is the place value of the 2 in 352.61?
   (A) tens (B) ones
   (C) tenths (D) hundreds

7. What is the place value of the 8 in 497.08?
   (A) tenths (B) hundreds
   (C) hundredths (D) tens

8. What is the place value of the 2 in 162.89?
   (A) ones (B) tens
   (C) tenths (D) hundreds

9. What is the place value of the 9 in 537.96?
   (A) tens (B) hundreds
   (C) tenths (D) hundredths

10. What is the place value of the 1 in 957.81?
    (A) hundreds (B) tens
    (C) tenths (D) hundredths

Practice 7

1. What digit is in the hundredths place in 246.75?
   (A) 5 (B) 6 (C) 4 (D) 2

2. What digit is in the hundredths place in 108.93?
   (A) 0 (B) 8 (C) 1 (D) 3

3. What digit is in the hundreds place in 650.81?
   (A) 0 (B) 8 (C) 5 (D) 6

4. What digit is in the tens place in 429.37?
   (A) 9 (B) 4 (C) 3 (D) 2

5. What digit is in the hundredths place in 62.94?
   (A) 2 (B) 4 (C) 0 (D) 6

6. What digit is in the ones place in 563.78?
   (A) 3 (B) 5 (C) 6 (D) 7

7. What digit is in the tenths place in 246.35?
   (A) 6 (B) 4 (C) 3 (D) 2

8. What digit is in the tenths place in 107.89?
   (A) 8 (B) 1 (C) 0 (D) 7

9. What digit is in the hundredths place in 650.71?
   (A) 5 (B) 1 (C) 6 (D) 0

10. What digit is in the hundreds place in 428.93?
    (A) 8 (B) 4 (C) 2 (D) 9

Practice 8

1. Write 46.49 in words.

(A) forty-six and forty-nine hundredths

(B) four thousand six hundred and forty-nine

(C) forty-six and forty-nine tenths

(D) forty-six point four and nine tenths

2. Write 83.97 in words.

(A) eighty-three and ninety-seven tenths

(B) eighty-three point ninety-seven hundredths

(C) eight thousand three hundred and ninety-seven

(D) eighty-three and ninety-seven hundredths

3. Write 94.28 in words.

(A) ninety-four and twenty-eight tenths

(B) nine thousand four hundred and twenty-eight

(C) ninety-four and twenty-eight hundredths

(D) ninety-four point twenty-eight hundredths

4. Write 52.91 in words.

(A) five thousand two hundred and ninety-one

(B) fifty-two point nine and one tenth

(C) fifty-two and ninety-one hundredths

(D) fifty-two and ninety-one tenths

5. Write 2.11 in words.

(A) two point eleven hundredths

(B) two hundred and eleven

(C) two and eleven hundredths

(D) two and eleven tenths

6. Write 31.92 in words.

(A) thirty-one point nine and two tenths

(B) three thousand one hundred and ninety-two

(C) thirty-one and ninety-two hundredths

(D) thirty-one and ninety-two tenths

7. Write 95.37 in words.

(A) ninety-five point thirty-seven hundredths

(B) ninety-five and thirty-seven hundredths

(C) ninety-five and thirty-seven tenths

(D) nine thousand five hundred and thirty-seven

8. Write 60.14 in words.

(A) sixty and fourteen tenths

(B) sixty point one and four tenths

(C) six thousand and fourteen

(D) sixty and fourteen hundredths

9. Write 24.98 in words.

(A) twenty-four point nine and eight tenths

(B) twenty-four and ninety-eight hundredths

(C) twenty-four and ninety-eight tenths

(D) two thousand four hundred and ninety-eight

10. Write 59.6 in words.

(A) fifty-nine point six tenths

(B) fifty-nine and six tenths

(C) fifty-nine and sixty tenths

(D) five thousand nine hundred and sixty

Practice 9

1. Which of the following is ordered from *least* to *greatest*?

(A)	(B)	(C)	(D)
0.1	2.6	4.5	0.1
3	0.1	3	2.6
2.6	3	2.6	3
4.5	4.5	0.1	4.5

2. Which of the following is ordered from *least* to *greatest*?

(A)	(B)	(C)	(D)
2	1.2	1	1
1.4	1	1.2	1.4
1.2	1.4	1.4	1.2
1	2	2	2

3. Which of the following is ordered from *least* to *greatest*?

(A)	(B)	(C)	(D)
7	1.2	1.1	1.1
1.6	1.1	1.2	1.6
1.2	1.6	1.6	1.2
1.1	7	7	7

4. Which of the following is ordered from *least* to *greatest*?

(A)	(B)	(C)	(D)
5	1.2	0.6	0.6
3	0.6	3	1.2
1.2	3	1.2	3
0.6	5	5	5

5. Which of the following is ordered from *least* to *greatest*?

(A)	(B)	(C)	(D)
0.9	2.6	0.5	0.5
0.5	1.4	1.4	0.9
1.4	0.9	0.9	1.4
2.6	0.5	2.6	2.6

6. Which of the following is ordered from *least* to *greatest*?

(A)	(B)	(C)	(D)
0.8	0.4	5	0.4
0.4	0.8	2.2	2.2
2.2	2.2	0.8	0.8
5	5	0.4	5

7. Which of the following is ordered from *least* to *greatest*?

(A)	(B)	(C)	(D)
0.8	0.3	2.5	0.3
0.3	0.8	1.2	1.2
1.2	1.2	0.8	0.8
2.5	2.5	0.3	2.5

Practice 10

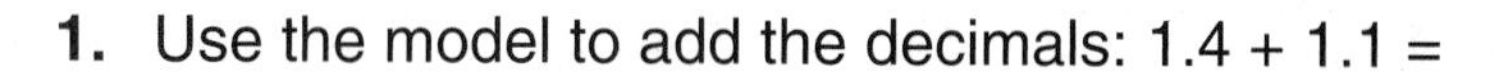

1. Use the model to add the decimals: 1.4 + 1.1 =

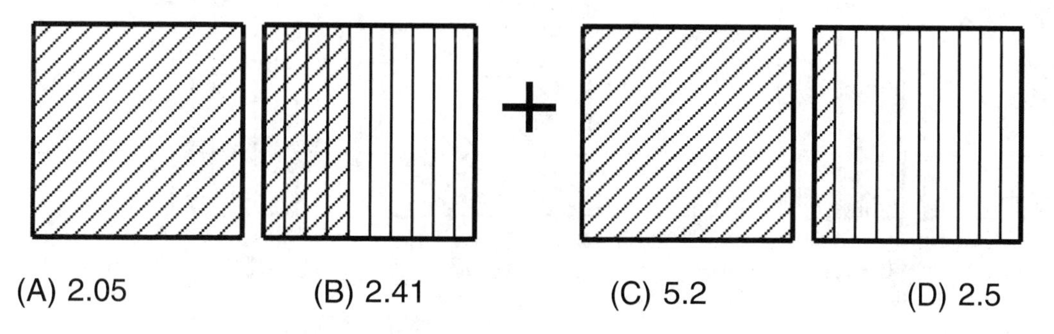

(A) 2.05 (B) 2.41 (C) 5.2 (D) 2.5

2. Use the model to add the decimals: 1.3 + 1.2 =

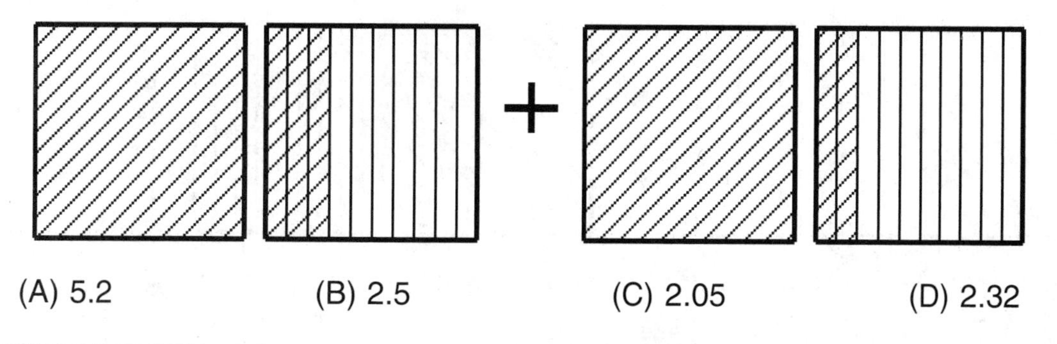

(A) 5.2 (B) 2.5 (C) 2.05 (D) 2.32

3. Use the model to add the decimals: 1.5 + 1.3 =

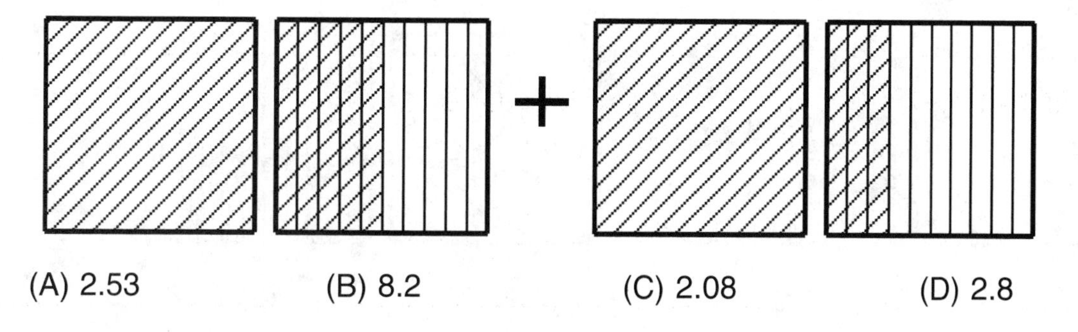

(A) 2.53 (B) 8.2 (C) 2.08 (D) 2.8

4. Use the model to add the decimals: 1.4 + 1.2 =

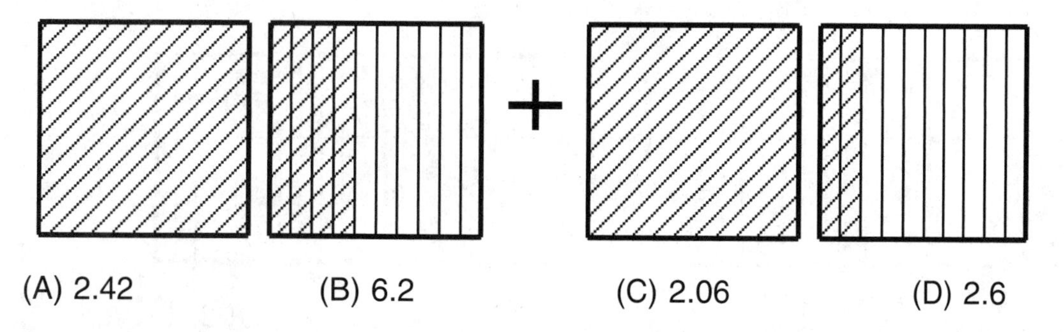

(A) 2.42 (B) 6.2 (C) 2.06 (D) 2.6

Practice 11

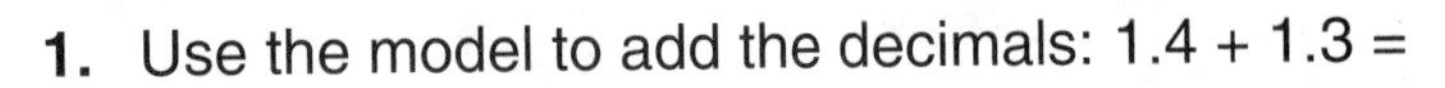

1. Use the model to add the decimals: 1.4 + 1.3 =

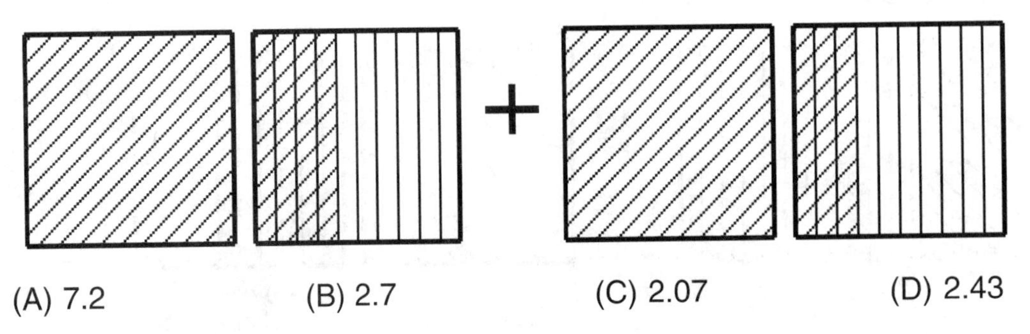

(A) 7.2 (B) 2.7 (C) 2.07 (D) 2.43

2. Use the model to add the decimals: 1.5 + 1.2 =

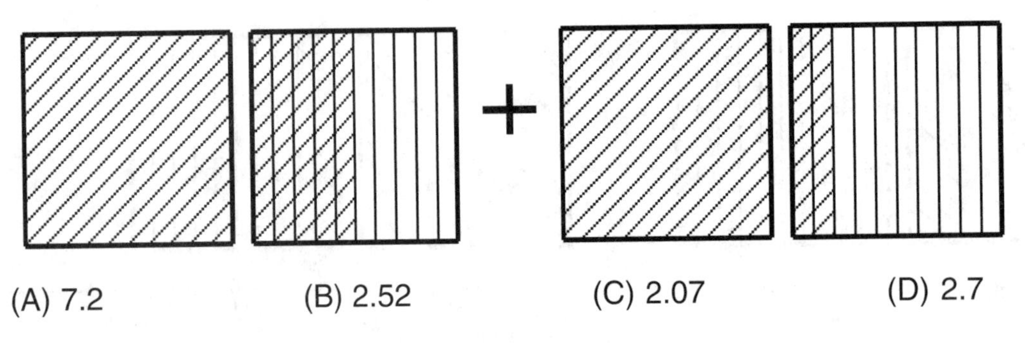

(A) 7.2 (B) 2.52 (C) 2.07 (D) 2.7

3. Use the model to add the decimals: 1.3 + 1.3 =

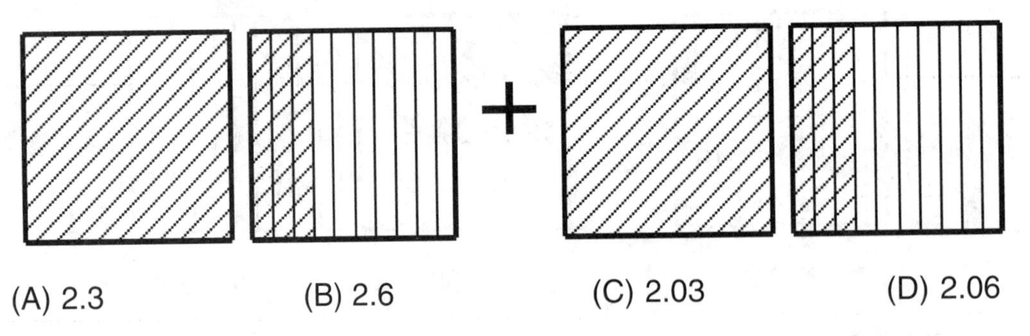

(A) 2.3 (B) 2.6 (C) 2.03 (D) 2.06

4. Use the model to add the decimals: 1.1 + 1.1 =

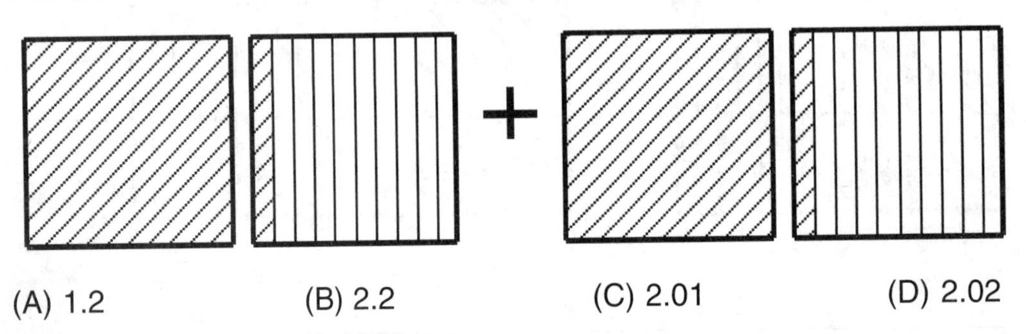

(A) 1.2 (B) 2.2 (C) 2.01 (D) 2.02

Practice 12

Solve by adding. Show your work.

1. $\begin{array}{r} 5.4 \\ +\ 3.1 \\ \hline \end{array}$	**8.** $\begin{array}{r} 7.4 \\ +\ 5.5 \\ \hline \end{array}$
2. $\begin{array}{r} 9.8 \\ +\ 3.0 \\ \hline \end{array}$	**9.** $\begin{array}{r} 4.2 \\ +\ 7.4 \\ \hline \end{array}$
3. $\begin{array}{r} 7.5 \\ +\ 3.9 \\ \hline \end{array}$	**10.** $\begin{array}{r} 2.6 \\ +\ 3.1 \\ \hline \end{array}$
4. $\begin{array}{r} 3.5 \\ +\ 5.7 \\ \hline \end{array}$	**11.** $\begin{array}{r} 7.4 \\ +\ 5.3 \\ \hline \end{array}$
5. $\begin{array}{r} 2.1 \\ +\ 5.5 \\ \hline \end{array}$	**12.** $\begin{array}{r} 5.0 \\ +\ 4.9 \\ \hline \end{array}$
6. $\begin{array}{r} 2.1 \\ +\ 9.2 \\ \hline \end{array}$	**13.** $\begin{array}{r} 4.6 \\ +\ 5.9 \\ \hline \end{array}$
7. $\begin{array}{r} 8.8 \\ +\ 5.2 \\ \hline \end{array}$	**14.** $\begin{array}{r} 2.9 \\ +\ 5.1 \\ \hline \end{array}$

Practice 13

Solve by adding. Show your work.

1.

$$\begin{array}{r} 6.21 \\ +\ .43 \\ \hline \end{array}$$

2.

$$\begin{array}{r} 7.05 \\ +\ .41 \\ \hline \end{array}$$

3.

$$\begin{array}{r} 18.01 \\ .5 \\ +\ 1.23 \\ \hline \end{array}$$

4.

$$\begin{array}{r} 24.3 \\ .7 \\ +\ 8.4 \\ \hline \end{array}$$

5.

$$\begin{array}{r} 16.3 \\ +\ 21.9 \\ \hline \end{array}$$

6.

$$\begin{array}{r} 82.1 \\ +\ .7 \\ \hline \end{array}$$

7.

$$\begin{array}{r} 2.27 \\ 3.41 \\ +\ 4.07 \\ \hline \end{array}$$

8.

$$\begin{array}{r} 1.2 \\ 6.3 \\ +\ 4.1 \\ \hline \end{array}$$

9.

$$\begin{array}{r} 22.9 \\ 7.2 \\ +\ .3 \\ \hline \end{array}$$

Practice 14

Solve by adding. Show your work.

1.	6.53 9.86 + 3.69	7.	59.28 89.84 + 11.28
2.	3.96 9.16 + 2.38	8.	7,854.27 3,855.22 + 3,905.38
3.	1.74 5.85 + 9.86	9.	3.55 9.21 + 9.94
4.	8.14 4.76 + 2.35	10.	7,469.91 6,551.25 + 1,563.47
5.	5.44 6.69 + 6.77	11.	8.31 7.54 + 4.11
6.	50.53 62.79 + 44.29	12.	3.69 6.27 + 6.21

Practice 15

Solve each problem. Show your work.

1.	8.0 13.52 + 11.31	**7.**	18.42 10.0 + 5.7
2.	6.0 11.2 + 12.17	**8.**	12.07 4.0 + 11.62
3.	14.3 10.0 +16.71	**9.**	8.5 17.06 + 9.31
4.	8.7 4.0 + 17.84	**10.**	13.05 6.0 + 12.7
5.	16.7 12.0 + 14.13	**11.**	41.50 8.03 + 7.00
6.	5.0 18.31 + 7.2	**12.**	1.34 4.72 + 1.08

Practice 16

1.

```
  9.7
- 8.9
```

(A) 1.9
(B) 0.8
(C) 1.8
(D) 0.9

2.

```
  6.1
- 2.4
```

(A) 2.7
(B) 3.7
(C) 2.6
(D) 3.6

3.

```
  4.6
- 2.8
```

(A) 1.9
(B) 2.8
(C) 2.9
(D) 1.8

4.

```
  7.3
- 2.1
```

(A) 5.2
(B) 5.3
(C) 6.2
(D) 6.3

5.

```
  3.9
- 1.7
```

(A) 2.2
(B) 2.1
(C) 1.1
(D) 1.2

6.

```
  8.8
- 5.2
```

(A) 2.5
(B) 3.5
(C) 3.6
(D) 2.6

7.

```
  5.4
- 3.3
```

(A) 2.2
(B) 3.2
(C) 2.1
(D) 3.1

8.

```
  2.2
- 1.6
```

(A) 1.6
(B) 0.6
(C) 0.7
(D) 1.7

9.

```
  8.6
- 2.7
```

(A) 4.9
(B) 4.8
(C) 5.9
(D) 5.8

10.

```
  5.8
- 2.4
```

(A) 3.3
(B) 2.3
(C) 2.4
(D) 3.4

Practice 17

Solve each problem. Show your work.

1. $\begin{array}{r} 14.76 \\ -\ 11.5 \\ \hline \end{array}$

2. $\begin{array}{r} 8.22 \\ -\ 3.2 \\ \hline \end{array}$

3. $\begin{array}{r} 6.31 \\ -\ 3.3 \\ \hline \end{array}$

4. $\begin{array}{r} 7.21 \\ -\ 2.1 \\ \hline \end{array}$

5. $\begin{array}{r} 10.28 \\ -\ 1.6 \\ \hline \end{array}$

6. $\begin{array}{r} 12.54 \\ -\ 5.7 \\ \hline \end{array}$

7. $\begin{array}{r} 13.75 \\ -\ 10.9 \\ \hline \end{array}$

8. $\begin{array}{r} 11.59 \\ -\ 1.4 \\ \hline \end{array}$

9. $\begin{array}{r} 5.37 \\ -\ 3.8 \\ \hline \end{array}$

10. $\begin{array}{r} 15.98 \\ -\ 5.9 \\ \hline \end{array}$

Practice 18

Solve by adding or subtracting. Show your work.

1. 25.62 .5 + 2.07	**5.** 71.53 .21 + 2.45	**9.** 15.7 − 2.5
2. 5.62 − 1.02	**6.** 1.9 11.25 + .27	**10.** 7.3 .5 + 4.6
3. 1.82 .07 + 3.1	**7.** 12.91 − 4.20	**11.** 12.45 − 10.2
4. 15.91 − 2.01	**8.** 8.1 + 2.9	**12.** 92.1 + 15.7

Practice 19

1. Round 68.54 to the nearest tenth.

   ______

2. Round 69.29 to the nearest tenth.

   ______

3. Round 91.58 to the nearest tenth.

   ______

4. Round 67.44 to the nearest tenth.

   ______

5. Round 75.764 to the nearest hundredth.

   ______

6. Round 36.319 to the nearest hundredth.

   ______

7. Round 83.34 to the nearest tenth.

   ______

8. Round 20.426 to the nearest hundredth.

   ______

9. Round 57.32 to the nearest tenth.

   ______

10. Round 8.832 to the nearest hundredth.

    ______

11. Round 49.347 to the nearest hundredth.

    ______

12. Round 67.16 to the nearest tenth.

    ______

Practice 20

1. What digits can you put in the box so that rounding 35.6☐ to the nearest tenth results in 35.7?

2. What digits can you put in the box so that rounding 44.1☐ to the nearest tenth results in 44.1?

3. What digits can you put in the box so that rounding 43.2☐ to the nearest tenth results in 43.3?

4. What digits can you put in the box so that rounding 40.5☐ to the nearest tenth results in 40.5?

5. What digits can you put in the box so that rounding 86.3☐ to the nearest tenth results in 86.4?

6. What digits can you put in the box so that rounding 30.8☐ to the nearest tenth results in 30.8?

7. What digits can you put in the box so that rounding 65.7☐ to the nearest tenth results in 65.8?

8. What digits can you put in the box so that rounding 29.4☐ to the nearest tenth results in 29.4?

9. What digits can you put in the box so that rounding 57.9☐ to the nearest tenth results in 58.0?

10. What digits can you put in the box so that rounding 27.8☐ to the nearest tenth results in 27.8?

Practice 21

1. Estimate by rounding each decimal to the nearest whole number:

   2.99 + 7.38

   (A) 11 (B) 9 (C) 10 (D) 8

2. Estimate by rounding each decimal to the nearest whole number:

   8.24 + 5.22

   (A) 13 (B) 12
   (C) 15 (D) 14

3. Estimate by rounding each decimal to the nearest whole number:

   7.12 + 6.35

   (A) 14 (B) 13
   (C) 12 (D) 15

4. Estimate by rounding each decimal to the nearest whole number:

   7.93 + 9.31

   (A) 16 (B) 17
   (C) 15 (D) 18

5. Estimate by rounding each decimal to the nearest whole number:

   3.22 + 4.03

   (A) 6 (B) 5 (C) 7 (D) 8

6. Estimate by rounding each decimal to the nearest whole number:

   2.1 + 7.75

   (A) 9 (B) 12 (C) 10 (D) 11

7. Estimate by rounding each decimal to the nearest whole number:

   8.47 + 4.8

   (A) 14 (B) 13
   (C) 12 (D) 15

8. Estimate by rounding each decimal to the nearest whole number:

   7.76 + 4.79

   (A) 14 (B) 15
   (C) 13 (D) 12

Practice 22          

1. Estimate by rounding each decimal to the nearest whole number:

$$\begin{array}{r} 50.49 \\ -\ \ 2.58 \\ \hline \end{array}$$

2. Estimate by rounding each decimal to the nearest whole number:

$$\begin{array}{r} 76.36 \\ -\ \ 4.68 \\ \hline \end{array}$$

3. Estimate by rounding each decimal to the nearest whole number:

$$\begin{array}{r} 88.72 \\ -\ \ 8.24 \\ \hline \end{array}$$

4. Estimate by rounding each decimal to the nearest whole number:

$$\begin{array}{r} 68.42 \\ -\ \ 2.16 \\ \hline \end{array}$$

5. Estimate by rounding each decimal to the nearest whole number:

$$\begin{array}{r} 28.68 \\ -\ \ 2.53 \\ \hline \end{array}$$

6. Estimate by rounding each decimal to the nearest whole number:

$$\begin{array}{r} 51.56 \\ -\ \ 5.88 \\ \hline \end{array}$$

7. Estimate by rounding each decimal to the nearest whole number:

$$\begin{array}{r} 44.45 \\ -\ \ 3.58 \\ \hline \end{array}$$

8. Estimate by rounding each decimal to the nearest whole number:

$$\begin{array}{r} 37.56 \\ -\ \ 4.34 \\ \hline \end{array}$$

Practice 23

1. What is the value of the change below? __________

2. What is the value of the change below? __________

3. What is the value of the change below? __________

4. What is the value of the change below? __________

Practice 24

1. Which shows the bills and coins you could use to pay exactly $7.68?
   (A) 2 five-dollar bills, 2 one-dollar bills, 2 quarters, 2 dimes, 1 nickel, 4 pennies
   (B) 2 five-dollar bills, 3 one-dollar bills, 3 quarters, 1 dime, 2 nickels, 3 pennies
   (C) 1 five-dollar bill, 2 one-dollar bills, 2 quarters, 1 dime, 1 nickel, 3 pennies
   (D) 1 five-dollar bill, 3 one-dollar bills, 2 quarters, 2 dimes, 2 nickels, 4 pennies

---

2. Which shows the bills and coins you could use to pay exactly $11.52?
   (A) 1 ten-dollar bill, 1 five-dollar bill, 1 one-dollar bill, 1 quarter, 3 dimes, 8 pennies
   (B) 2 one-dollar bills, 1 quarter, 3 dimes, 8 pennies
   (C) 1 ten-dollar bill, 1 one-dollar bill, 1 quarter, 2 dimes, 7 pennies
   (D) 1 five-dollar bill, 2 one-dollar bills, 2 quarters, 2 dimes, 7 pennies

---

3. Which shows the bills and coins you could use to pay exactly $15.88?
   (A) 2 five-dollar bills, 1 one-dollar bill, 4 quarters, 2 nickels, 8 pennies
   (B) 1 ten-dollar bill, 2 five-dollar bills, 3 quarters, 1 dime, 1 nickel, 9 pennies
   (C) 1 five-dollar bill, 1 one-dollar bill, 3 quarters, 1 dime, 2 nickels, 9 pennies
   (D) 1 ten-dollar bill, 1 five-dollar bill, 3 quarters, 1 nickel, 8 pennies

---

4. Which shows the bills and coins you could use to pay exactly $4.26?
   (A) 4 one-dollar bills, 2 dimes, 6 pennies
   (B) 1 five-dollar bill, 5 one-dollar bills, 2 dimes, 6 pennies
   (C) 1 five-dollar bill, 4 one-dollar bills, 3 dimes, 7 pennies
   (D) 5 one-dollar bills, 3 dimes, 7 pennies

---

5. Which shows the bills and coins you could use to pay exactly $8.80?
   (A) 1 five-dollar bill, 3 one-dollar bills, 3 quarters, 5 pennies
   (B) 2 five-dollar bills, 3 one-dollar bills, 3 quarters, 1 dime, 6 pennies
   (C) 2 five-dollar bills, 4 one-dollar bills, 4 quarters, 5 pennies
   (D) 1 five-dollar bill, 4 one-dollar bills, 3 quarters, 1 dime, 6 pennies

---

6. Which shows the bills and coins you could use to pay exactly $14.19?
   (A) 5 one-dollar bills, 2 dimes, 2 nickels, 5 pennies
   (B) 1 five-dollar bill, 5 one-dollar bills, 1 dime, 2 nickels, 4 pennies
   (C) 1 ten-dollar bill, 1 five-dollar bill, 4 one-dollar bills, 2 dimes, 1 nickel, 5 pennies
   (D) 1 ten-dollar bill, 4 one-dollar bills, 1 dime, 1 nickel, 4 pennies

Practice 25

1. List the bills and coins you could use to pay the exact amount: $12.69
   (A) one five; three ones; three quarters; one dime; two nickels; four pennies
   (B) three ones; two quarters; two dimes; two nickels; five pennies
   (C) one ten; one five; two ones; two quarters; two dimes; one nickel; five pennies
   (D) one ten; two ones; two quarters; one dime; one nickel; four pennies

2. List the bills and coins you could use to pay the exact amount: $8.30
   (A) two fives; three ones; one quarter; one dime; six pennies
   (B) one five; three ones; one quarter; five pennies
   (C) two fives; four ones; two quarters; five pennies
   (D) one five; four ones; one quarter; one dime; six pennies

3. List the bills and coins you could use to pay the exact amount: $9.26
   (A) one five; five ones; three dimes; seven pennies
   (B) two fives; five ones; two dimes; six pennies
   (C) two fives; four ones; three dimes; seven pennies
   (D) one five; four ones; two dimes; six pennies

4. List the bills and coins you could use to pay the exact amount: $10.98
   (A) one ten; one five; three quarters; two dimes; one nickel; nine pennies
   (B) one one; three quarters; two dimes; two nickels; nine pennies
   (C) one ten; three quarters; one dime; one nickel; eight pennies
   (D) one five; one one; four quarters; one dime; two nickels; eight pennies

5. List the bills and coins you could use to pay the exact amount: $6.23
   (A) two fives; one one; three dimes; four pennies
   (B) one five; one one; two dimes; three pennies
   (C) one five; two ones; three dimes; four pennies
   (D) two fives; two ones; two dimes; three pennies

Practice 26

1. How many nickels does it take to make 80¢? (A) 15 (B) 32 (C) 16 (D) 17

2. How many dimes does it take to make 80¢? (A) 9 (B) 16 (C) 8 (D) 7

3. How many quarters does it take to make 50¢? (A) 3 (B) 2 (C) 1 (D) 4

4. How many dimes does it take to make 70¢? (A) 8 (B) 14 (C) 6 (D) 7

5. How many nickels does it take to make 35¢? (A) 7 (B) 8 (C) 14 (D) 6

6. You have six quarters, four dimes and five nickels. You can choose 1 item to buy. What are the toys that you can choose?

Yo-Yo	Modeling Clay	Paint Set	Board Game	Puzzle	Jump Rope
$2.50	$1.50	$1.00	$4.75	$5.00	$3.00

7. You bought a poster that cost $4.37. List the bills and coins that you could receive as change from a five-dollar bill.

8. You bought a book that cost $4.48. List the bills and coins that you could receive as change from a five-dollar bill.

9. You bought a hat that cost $4.93. List the bills and coins that you could receive as change from a five-dollar bill.

10. You bought a game that cost $4.12. List the bills and coins that you could receive as change from a five-dollar bill.

Practice 27

1. $21.34 + $4.32
   (A) $25.02 (B) $23.02 (C) $25.66 (D) $33.06

2. $43.01 + $3.43
   (A) $40.42 (B) $46.42 (C) $46.44 (D) $56.44

3. $34.20 + $1.13
   (A) $33.13 (B) $35.13 (C) $35.33 (D) $43.13

4. $42.10 + $1.22
   (A) $41.12 (B) $43.12 (C) $51.12 (D) $43.32

5. $22.34 + $2.20
   (A) $24.54 (B) $24.14 (C) $30.14 (D) $20.14

6. $13.23 + $3.40
   (A) $26.23 (B) $16.23 (C) $10.23 (D) $16.63

7. $31.12 + $2.34
   (A) $41.26 (B) $31.22 (C) $33.22 (D) $33.46

8. $24.03 + $1.11
   (A) $23.12 (B) $25.12 (C) $35.14 (D) $25.14

9. $42.11 + $3.34
   (A) $41.23 (B) $45.23 (C) $55.25 (D) $45.45

10. $13.42 + $4.13
    (A) $11.31 (B) $17.31 (C) $21.35 (D) $17.55

Practice 28

1. $34.74 + $3.29 =
   (A) $31.55 (B) $38.03 (C) $37.55 (D) $47.63

2. $12.38 + $2.44 =
   (A) $14.82 (B) $10.14 (C) $14.14 (D) $20.22

3. $81.67 + $6.54 =
   (A) $85.13 (B) $95.21 (C) $88.21 (D) $87.13

4. $69.23 + $7.39 =
   (A) $76.62 (B) $62.16 (C) $76.16 (D) $86.22

5. $97.82 + $8.98 =
   (A) $91.16 (B) $105.16 (C) $106.80 (D) $101.20

6. $25.19 + $4.64 =
   (A) $29.83 (B) $39.63 (C) $29.55 (D) $21.55

7. $46.91 + $9.89 =
   (A) $43.18 (B) $56.80 (C) $65.20 (D) $55.18

8. $53.56 + $1.18 =
   (A) $62.54 (B) $54.42 (C) $52.42 (D) $54.74

9. $78.45 + $5.75 =
   (A) $84.20 (B) $73.30 (C) $83.40 (D) $83.30

10. $61.59 + $2.31 =
    (A) $63.90 (B) $61.28 (C) $73.30 (D) $63.28

Practice 29

Solve each problem. Show your work.

1.
```
  $7.35
   2.00
+  0.22
-------
```

6.
```
  $4.20
   2.15
+  0.45
-------
```

2.
```
  $6.20
   2.00
+  0.45
-------
```

7.
```
  $8.45
   7.00
+  0.97
-------
```

3.
```
  $8.45
   7.10
+  2.00
-------
```

8.
```
  $8.72
   3.00
+  0.72
-------
```

4.
```
  $4.95
   2.00
+  0.70
-------
```

9.
```
  $3.27
   0.03
+  7.00
-------
```

5.
```
  $6.47
   0.54
+  2.00
-------
```

10.
```
  $5.33
   0.07
+  3.00
-------
```

Practice 30

Solve each problem. Show your work.

1. $5.45
 − 0.82

7. $6.30
 − 0.58

2. $5.10
 − 0.30

8. $5.90
 − 0.61

3. $6.85
 − 0.75

9. $5.85
 − 0.25

4. $6.70
 − 0.49

10. $6.30
 − 0.35

5. $5.65
 − 0.32

11. $6.60
 − 0.87

6. $6.45
 − 0.50

12. $5.70
 − 0.46

Practice 31

Solve each problem. Show your work.

1. $65.85 − 0.68	7. $69.75 − 0.37
2. $85.61 − 0.77	8. $52.95 − 0.47
3. $89.62 − 0.43	9. $99.55 − 0.56
4. $75.14 − 0.56	10. $57.95 − 0.38
5. $52.52 − 0.76	11. $74.22 − 0.73
6. $65.34 − 0.57	12. $97.64 − 0.89

Practice 32

1. Estimate by rounding to the nearest dollar: \$2.12 + \$5.23
   (A) \$8 (B) \$7 (C) \$31 (D) \$5

2. Estimate by rounding to the nearest dollar: \$9.78 + \$7.31
   (A) \$17 (B) \$25 (C) \$19 (D) \$18

3. Estimate by rounding to the nearest dollar: \$6.32 + \$9.87
   (A) \$36 (B) \$18 (C) \$16 (D) \$15

4. Estimate by rounding to the nearest dollar: \$7.31 + \$1.27
   (A) \$7 (B) \$6 (C) \$60 (D) \$8

5. Estimate by rounding to the nearest dollar: \$8.87 + \$1.12
   (A) \$12 (B) \$78 (C) \$10 (D) \$9

6. Estimate by rounding to the nearest dollar: \$6.71 + \$3.98
   (A) \$12 (B) \$27 (C) \$9 (D) \$11

7. Estimate by rounding to the nearest dollar: \$3.11 + \$1.83
   (A) \$13 (B) \$5 (C) \$4 (D) \$7

8. Estimate by rounding to the nearest dollar: \$4.82 + \$2.11
   (A) \$27 (B) \$5 (C) \$7 (D) \$8

9. Estimate by rounding to the nearest dollar: \$1.11 + \$3.23
   (A) \$21 (B) \$2 (C) \$3 (D) \$4

10. Estimate by rounding to the nearest dollar: \$1.82 + \$3.31
    (A) \$5 (B) \$15 (C) \$4 (D) \$7

Practice 33

1. Estimate by rounding to the nearest dollar: $9.24
   − 3.53

   (A) $5 (B) $4 (C) $12 (D) $6

2. Estimate by rounding to the nearest dollar: $8.42
   − 4.97

   (A) $4 (B) $12 (C) $2 (D) $3

3. Estimate by rounding to the nearest dollar: $6.14
   − 1.68

   (A) $3 (B) $4 (C) $7 (D) $5

4. Estimate by rounding to the nearest dollar: $7.10
   − 2.67

   (A) $7 (B) $4 (C) $5 (D) $9

5. Estimate by rounding to the nearest dollar: $8.33
   − 3.78

   (A) $4 (B) $11 (C) $3 (D) $5

6. Estimate by rounding to the nearest dollar: $6.06
   − 4.54

   (A) $2 (B) $4 (C) $1 (D) $10

7. Estimate by rounding to the nearest dollar: $7.18
   − 2.79

   (A) $4 (B) $5 (C) $3 (D) $9

8. Estimate by rounding to the nearest dollar: $8.08
   − 4.98

   (A) $12 (B) $4 (C) $6 (D) $3

9. Estimate by rounding to the nearest dollar: $7.25
   − 2.71

   (A) $9 (B) $4 (C) $7 (D) $5

10. Estimate by rounding to the nearest dollar: $6.44
    − 3.60

    (A) $9 (B) $5 (C) $2 (D) $3

11. Estimate by rounding to the nearest dollar: $6.23
    − 4.63

    (A) $1 (B) $10 (C) $2 (D) $4

12. Estimate by rounding to the nearest dollar: $9.36
    − 3.70

    (A) $8 (B) $5 (C) $6 (D) $12

13. Estimate by rounding to the nearest dollar: $7.34
    − 2.92

    (A) $5 (B) $4 (C) $3 (D) $9

14. Estimate by rounding to the nearest dollar: $6.45
    − 2.64

    (A) $6 (B) $8 (C) $3 (D) $4

Practice 34

1. $71.11 × 6 = (A) $373.26 (B) $433.26 (C) $366.66 (D) $426.66

2. $34.64 × 7 = (A) $242.48 (B) $164.78 (C) $172.48 (D) $234.78

3. $47.89 × 4 = (A) $231.56 (B) $195.96 (C) $191.56 (D) $235.96

4. $29.42 × 2 = (A) $58.84 (B) $56.64 (C) $38.84 (D) $36.64

5. $83.28 × 3 = (A) $219.84 (B) $223.14 (C) $253.14 (D) $249.84

6. $55.73 × 9 = (A) $581.67 (B) $491.67 (C) $591.57 (D) $501.57

7. $71.57 × 7 = (A) $500.99 (B) $508.69 (C) $430.99 (D) $438.69

8. $34.21 × 5 = (A) $221.05 (B) $171.05 (C) $215.55 (D) $165.55

9. $49.13 × 2 = (A) $80.46 (B) $98.26 (C) $78.26 (D) $100.46

10. $27.89 × 8 = (A) $223.12 (B) $294.32 (C) $303.12 (D) $214.32

Practice 35

		(A)	(B)	(C)	(D)
1.	$36.14 × 3	(A) $108.42	(B) $78.42	(C) $105.42	(D) $108.12
2.	$69.43 × 5	(A) $352.15	(B) $347.65	(C) $397.15	(D) $347.15
3.	$93.36 × 7	(A) $583.52	(B) $653.52	(C) $652.82	(D) $646.52
4.	$74.67 × 8	(A) $597.36	(B) $598.16	(C) $677.36	(D) $605.36
5.	$55.29 × 9	(A) $587.61	(B) $506.61	(C) $497.61	(D) $498.51
6.	$28.58 × 6	(A) $165.48	(B) $170.88	(C) $111.48	(D) $171.48
7.	$87.72 × 2	(A) $175.44	(B) $175.64	(C) $177.44	(D) $195.44
8.	$42.91 × 4	(A) $171.64	(B) $171.24	(C) $167.64	(D) $131.64
9.	$11.85 × 8	(A) $94.80	(B) $14.80	(C) $94.00	(D) $86.80
10.	$56.44 × 6	(A) $338.04	(B) $278.64	(C) $332.64	(D) $338.64

Practice 36

Solve by multiplying. Show your work.

1. $17.35 x 5	8. $41.59 x 9
2. $34.48 x 6	9. $31.34 x 2
3. $62.86 x 5	10. $37.44 x 5
4. $74.95 x 5	11. $43.36 x 3
5. $52.85 x 6	12. $13.63 x 7
6. $36.24 x 7	13. $75.12 x 9
7. $43.34 x 5	14. $11.45 x 3

Use with Answer Sheet (page 46)

Test Practice 1

1. What decimal matches the *shaded* part of this rectangle?

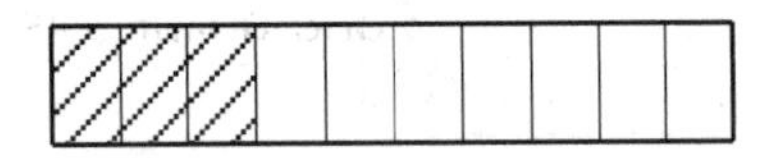

(A) 3 (B) 0.3 (C) 0.03 (D) 3.7

2. What decimal matches the *shaded* part of these rectangles?

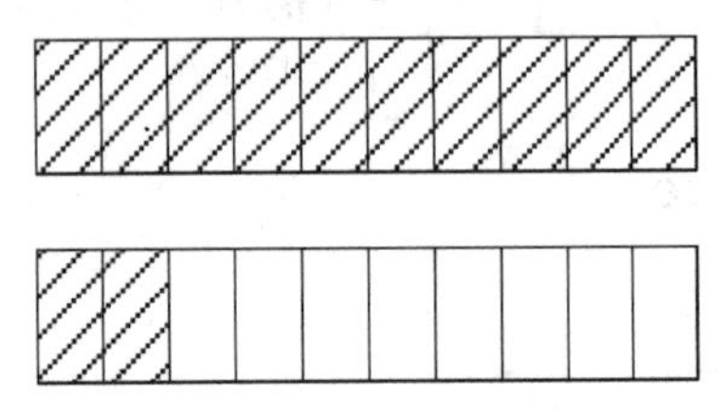

(A) 1.2 (B) 0.12
(C) 1.02 (D) 12.0

3. What is the place value of the 2 in 927.05?

(A) ones (B) tens
(C) hundreds (D) tenths

4. What digit is in the tenths place in 465.73?

(A) 5 (B) 4 (C) 6 (D) 7

5. Which of the following is ordered from *least* to *greatest*?

(A)	(B)	(C)	(D)
4	1	0.4	0.4
2.4	0.4	2.4	1
1	2.4	1	2.4
0.4	4	4	4

6. What is the place value of the 3 in 209.35?

(A) tens (B) tenths
(C) hundredths (D) hundreds

7. Which of the following is ordered from *least* to *greatest*?

(A)	(B)	(C)	(D)
1.2	0.4	0.4	2.6
0.4	1.5	1.2	1.5
1.5	1.2	1.5	1.2
2.6	2.6	2.6	0.4

8. What digit is in the tenths place in 891.02?

(A) 9 (B) 1 (C) 8 (D) 0

9. What digit is in the ones place in 706.58?

(A) 7 (B) 6 (C) 5 (D) 0

10. What is the place value of the 8 in 413.68?

(A) hundredths (B) hundreds
(C) tens (D) tenths

11. What decimal matches the *shaded* part of this rectangle?

(A) 0.08 (B) 0.8 (C) 8.2 (D) 8

12. What decimal matches the *shaded* part of these rectangles?

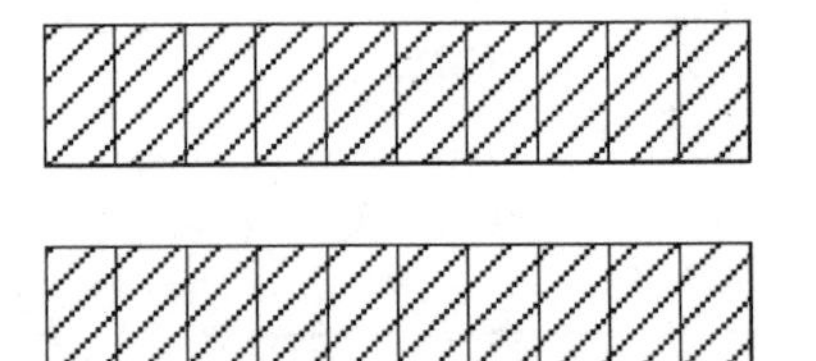

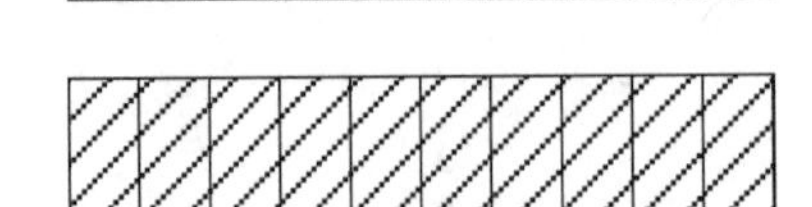

(A) 3.6 (B) 36.0
(C) 0.36 (D) 3.06

13. What is the place value of the 6 in 617.95?

(A) tens (B) hundredths
(C) tenths (D) hundreds

Test Practice 2

1. What is the value of the change below?

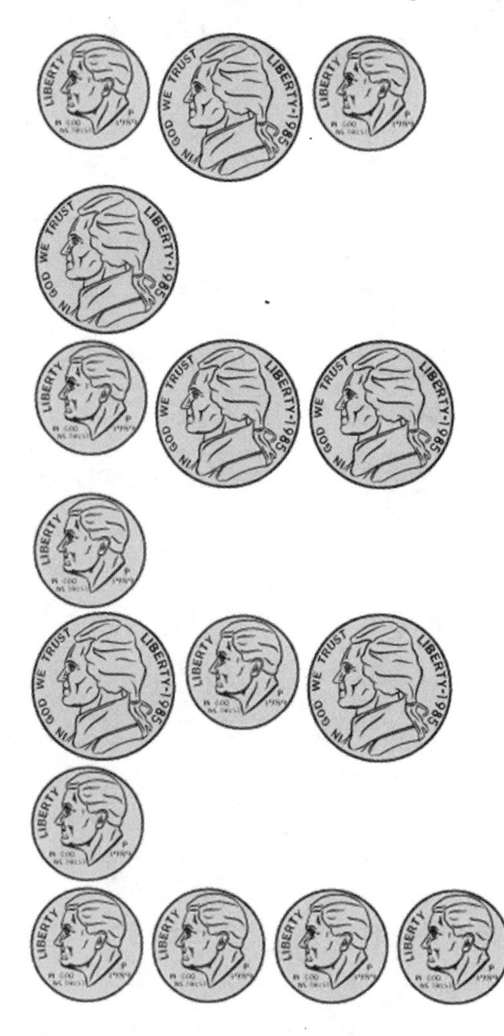

(A) $1.40 (B) $1.10
(C) $1.30 (D) $1.90

2. Which shows the bills and coins you could use to pay exactly $16.21?
   (A) 1 ten-dollar bill, 1 five-dollar bill, 1 one-dollar bill, 1 dime, 1 nickel, 6 pennies
   (B) 1 ten-dollar bill, 2 five-dollar bills, 1 one-dollar bill, 2 dimes, 1 nickel, 7 pennies
   (C) 1 five-dollar bill, 2 one-dollar bills, 2 dimes, 2 nickels, 7 pennies
   (D) 2 five-dollar bills, 2 one-dollar bills, 1 dime, 2 nickels, 6 pennies

3. Which shows the bills and coins you could use to pay exactly $2.29?
   (A) 1 five-dollar bill, 3 one-dollar bills, 2 quarters, 4 pennies
   (B) 3 one-dollar bills, 1 quarter, 1 dime, 5 pennies
   (C) 2 one-dollar bills, 1 quarter, 4 pennies
   (D) 1 five-dollar bill, 2 one-dollar bills, 1 quarter, 1 dime, 5 pennies

4. Write a decimal for the *shaded* part of this box.

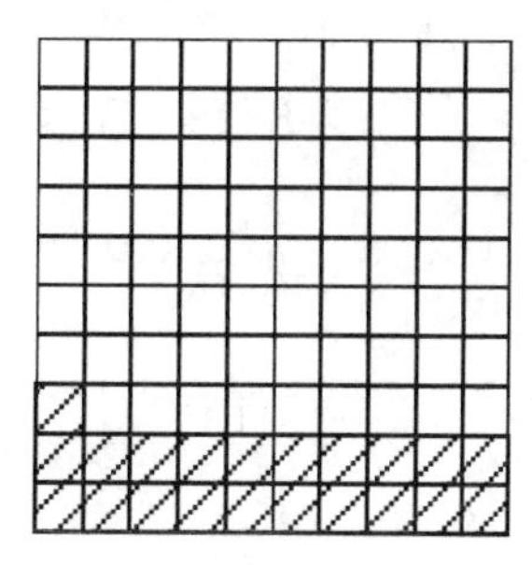

(A) 0.21 (B) 0.80
(C) 0.20 (D) 0.79

5. Write 18.42 in words.
   (A) one thousand eight hundred and forty-two
   (B) eighteen and forty-two hundredths
   (C) eighteen and forty-two tenths
   (D) eighteen point four and two tenths

6. Write 5.86 in words.
   (A) five and eighty-six tenths
   (B) five point eighty-six hundredths
   (C) five hundred and eighty-six
   (D) five and eighty-six hundredths

7.
$$\begin{array}{r} 2.63 \\ 1.53 \\ +\ 4.14 \\ \hline \end{array}$$

(A) 7.19 (B) 8.30
(C) 8.29 (D) 8.20

Test Practice 3

1. List the bills and coins you could use to pay the exact amount: $5.68
   - (A) one five; two quarters; one dime; eight pennies
   - (B) two fives; two quarters; two dimes; nine pennies
   - (C) two fives; one one; three quarters; one dime; eight pennies
   - (D) one five; one one; two quarters; two dimes; nine pennies

2. Estimate by rounding each decimal to the nearest whole number: 3.13 + 1.48

   (A) 6 (B) 3 (C) 5 (D) 4

3. List the bills and coins you could use to pay the exact amount: $13.08
   - (A) one ten; one five; three ones; one dime; one nickel; four pennies
   - (B) four ones; one dime; two nickels; four pennies
   - (C) one five; four ones; two nickels; three pennies
   - (D) one ten; three ones; one nickel; three pennies

4. Round 16.48 to the nearest tenth.

   (A) 16.5 (B) 16.4
   (C) 16.6 (D) 16

5. How many nickels does it take to make 50¢?

   (A) 10 (B) 9 (C) 20 (D) 11

6. $5.50 − 0.42

   (A) $5.08 (B) $4.93
   (C) $5.92 (D) $6.08

7. How many nickels does it take to make 40¢?

   (A) 9 (B) 8 (C) 16 (D) 7

8. $44.69 × 2 =

   (A) $89.38 (B) $69.38
   (C) $67.18 (D) $87.18

9. $21.18 × 5

   (A) $105.90 (B) $106.40
   (C) $110.90 (D) $155.90

10. $87.27 × 6 =

    (A) $470.22 (B) $530.22
    (C) $523.62 (D) $463.62

11. $5.45 − 0.40

    (A) $5.85 (B) $6.05
    (C) $4.90 (D) $5.05

12. 6.25 + 7.82 + 4.51

    (A) 17.67 (B) 18.48
    (C) 18.57 (D) 18.58

13. Estimate by rounding each decimal to the nearest whole number: 1.64 + 2.93

    (A) 6 (B) 4 (C) 5 (D) 3

14. Which numbers below round to 16 when rounded to the nearest whole number?

    16.27 15.47 16.63 15.15
    15.89 16.49 15.63 16.79

    (A) 16.27, 16.49, 16.63, 16.79
    (B) 15.15, 15.47, 16.63, 15.63
    (C) 15.89, 16.63, 15.15, 16.79
    (D) 16.27, 15.63, 16.49, 15.89

15. Estimate by rounding each decimal to the nearest whole number: 4.18 + 9.64

    (A) 14 (B) 12 (C) 13 (D) 15

Test Practice 4

Directions: Fill in the circle for the correct answer to each addition problem. Choose "none of these" if the right answer is not given.

1.

$ 5.63
\+ 4.17

(A) $9.86
(B) $9.70
(C) $9.46
(D) $9.80
(E) none of these

2.

$ 3.48
\+ 1.52

(A) $4.90
(B) $4.96
(C) $5.50
(D) $5.00
(E) none of these

3.

$ 24.18
\+ 16.53

(A) $40.71
(B) $41.71
(C) $40.17
(D) $40.61
(E) none of these

4.

37.2 + 6.4 + 5.3 =

(A) 4.89
(B) 489
(C) 38.9
(D) 48.9
(E) none of these

5.

$ 22.99
\+ 7.13

(A) $301.2
(B) $30.12
(C) $30.02
(D) $34.12
(E) none of these

6.

4.2 + 3.8 =

(A) 8.1
(B) 8.6
(C) 7.4
(D) 8.4
(E) none of these

7. Dustin ran half of a mile in gym class in 129.6 seconds. If he ran one mile at the same speed, how many seconds would it take?

(A) 300.9
(B) 258.9
(C) 250.9
(D) 259.2
(E) none of these

8. If you had $11.33 saved for a present that cost $18.80, how much more money would you need?

(A) $9.47
(B) $7.47
(C) $18.80
(D) $7.53
(E) none of these

9.

36.3 + 50.10 =

(A) 86.04
(B) 43.61
(C) 53.73
(D) 86.43
(E) none of these

10.

$ 39.99
\+ 8.99

(A) $48.98
(B) $489.80
(C) $48.00
(D) $31.00
(E) none of these

Test Practice 5

Directions: Fill in the circle for the correct answer to each subtraction problem. Choose "none of these" if the right answer is not given.

1.

$$\begin{array}{r} \$\,67.40 \\ -\ 14.39 \\ \hline \end{array}$$

(A) $53.19
(B) $53.79
(C) $53.01
(D) $81.79
(E) none of these

2.

$$\begin{array}{r} \$\,14.01 \\ -\ 7.17 \\ \hline \end{array}$$

(A) $6.84
(B) $68.40
(C) $21.18
(D) $7.04
(E) none of these

3.

$$\begin{array}{r} 400.00 \\ -\ 0.7 \\ \hline \end{array}$$

(A) 40.00
(B) 399.3
(C) 399.7
(D) 370
(E) none of these

4.

6.3 – 2.8 =

(A) 623.8
(B) 9.1
(C) 4.5
(D) .35
(E) none of these

5.

9.23 – 8.47 =

(A) 76
(B) 17.70
(C) 7.6
(D) 0.76
(E) none of these

6.

0.38 – 0.21 =

(A) 1.7
(B) 1.07
(C) .59
(D) 5.9
(E) none of these

7.

8.5 – 3.9 =

(A) 4.7
(B) 12.5
(C) 12.4
(D) 5.3
(E) none of these

8.

$$\begin{array}{r} \$383.00 \\ -\ 48.53 \\ \hline \end{array}$$

(A) $431.53
(B) $334.53
(C) $334.47
(D) $335.47
(E) none of these

9.

12.8 – 3.8 =

(A) 9.0
(B) 15.6
(C) 90
(D) 15.0
(E) none of these

10.

$$\begin{array}{r} \$63.41 \\ -\ 28.05 \\ \hline \end{array}$$

(A) $91.46
(B) $35.36
(C) $353.60
(D) $3,536.00
(E) none of these

Test Practice 6

1. Which is four dollars and twenty-seven cents?

   Ⓐ 4.27¢
   Ⓑ $427
   Ⓒ $40027
   Ⓓ $4.27

2. How do you write $3.12?

   Ⓐ three twelve
   Ⓑ three dollars and twelve cents
   Ⓒ three point twelve cents
   Ⓓ dollars three and twelve

3. Mark cuts the grass for one dollar an hour. He worked from two o'clock to five o'clock. How much did he earn?

   Ⓐ $4.00
   Ⓑ $2.00
   Ⓒ $3.00
   Ⓓ $7.00

4. Lupita baby-sits for two dollars an hour. She baby-sat from 6:00 to 8:30. How much did she earn?

   Ⓐ $4.00
   Ⓑ $4.50
   Ⓒ $28.00
   Ⓓ $5.00

5. Which group is closest in amount to one dollar?

Ⓐ

Ⓑ

Ⓒ

Ⓓ

6. Pretend this is a subtraction problem. Fill the circle next to your answer.

______

   Ⓐ 1 penny
   Ⓑ 4 pennies
   Ⓒ 1 dime
   Ⓓ 1 nickel

7. Pretend this a subtraction problem. Fill the circle next to your answer.

–

=

______

   Ⓐ 3 dimes, 1 nickel
   Ⓑ 4 dimes
   Ⓒ 4 dimes, 1 nickel
   Ⓓ 3 dimes

Answer Sheet

Test Practice 1

1. Ⓐ Ⓑ Ⓒ Ⓓ
2. Ⓐ Ⓑ Ⓒ Ⓓ
3. Ⓐ Ⓑ Ⓒ Ⓓ
4. Ⓐ Ⓑ Ⓒ Ⓓ
5. Ⓐ Ⓑ Ⓒ Ⓓ
6. Ⓐ Ⓑ Ⓒ Ⓓ
7. Ⓐ Ⓑ Ⓒ Ⓓ
8. Ⓐ Ⓑ Ⓒ Ⓓ
9. Ⓐ Ⓑ Ⓒ Ⓓ
10. Ⓐ Ⓑ Ⓒ Ⓓ
11. Ⓐ Ⓑ Ⓒ Ⓓ
12. Ⓐ Ⓑ Ⓒ Ⓓ
13. Ⓐ Ⓑ Ⓒ Ⓓ

Test Practice 2

1. Ⓐ Ⓑ Ⓒ Ⓓ
2. Ⓐ Ⓑ Ⓒ Ⓓ
3. Ⓐ Ⓑ Ⓒ Ⓓ
4. Ⓐ Ⓑ Ⓒ Ⓓ
5. Ⓐ Ⓑ Ⓒ Ⓓ
6. Ⓐ Ⓑ Ⓒ Ⓓ
7. Ⓐ Ⓑ Ⓒ Ⓓ

Test Practice 3

1. Ⓐ Ⓑ Ⓒ Ⓓ
2. Ⓐ Ⓑ Ⓒ Ⓓ
3. Ⓐ Ⓑ Ⓒ Ⓓ
4. Ⓐ Ⓑ Ⓒ Ⓓ
5. Ⓐ Ⓑ Ⓒ Ⓓ
6. Ⓐ Ⓑ Ⓒ Ⓓ
7. Ⓐ Ⓑ Ⓒ Ⓓ
8. Ⓐ Ⓑ Ⓒ Ⓓ
9. Ⓐ Ⓑ Ⓒ Ⓓ
10. Ⓐ Ⓑ Ⓒ Ⓓ
11. Ⓐ Ⓑ Ⓒ Ⓓ
12. Ⓐ Ⓑ Ⓒ Ⓓ
13. Ⓐ Ⓑ Ⓒ Ⓓ
14. Ⓐ Ⓑ Ⓒ Ⓓ
15. Ⓐ Ⓑ Ⓒ Ⓓ

Test Practice 4

1. Ⓐ Ⓑ Ⓒ Ⓓ Ⓔ
2. Ⓐ Ⓑ Ⓒ Ⓓ Ⓔ
3. Ⓐ Ⓑ Ⓒ Ⓓ Ⓔ
4. Ⓐ Ⓑ Ⓒ Ⓓ Ⓔ
5. Ⓐ Ⓑ Ⓒ Ⓓ Ⓔ
6. Ⓐ Ⓑ Ⓒ Ⓓ Ⓔ
7. Ⓐ Ⓑ Ⓒ Ⓓ Ⓔ
8. Ⓐ Ⓑ Ⓒ Ⓓ Ⓔ
9. Ⓐ Ⓑ Ⓒ Ⓓ Ⓔ
10. Ⓐ Ⓑ Ⓒ Ⓓ Ⓔ

Test Practice 5

1. Ⓐ Ⓑ Ⓒ Ⓓ Ⓔ
2. Ⓐ Ⓑ Ⓒ Ⓓ Ⓔ
3. Ⓐ Ⓑ Ⓒ Ⓓ Ⓔ
4. Ⓐ Ⓑ Ⓒ Ⓓ Ⓔ
5. Ⓐ Ⓑ Ⓒ Ⓓ Ⓔ
6. Ⓐ Ⓑ Ⓒ Ⓓ Ⓔ
7. Ⓐ Ⓑ Ⓒ Ⓓ Ⓔ
8. Ⓐ Ⓑ Ⓒ Ⓓ Ⓔ
9. Ⓐ Ⓑ Ⓒ Ⓓ Ⓔ
10. Ⓐ Ⓑ Ⓒ Ⓓ Ⓔ

Test Practice 6

1. Ⓐ Ⓑ Ⓒ Ⓓ
2. Ⓐ Ⓑ Ⓒ Ⓓ
3. Ⓐ Ⓑ Ⓒ Ⓓ
4. Ⓐ Ⓑ Ⓒ Ⓓ
5. Ⓐ Ⓑ Ⓒ Ⓓ
6. Ⓐ Ⓑ Ⓒ Ⓓ
7. Ⓐ Ⓑ Ⓒ Ⓓ

Answer Key

Page 4
1. A
2. B
3. C
4. D
5. C
6. C
7. B
8. A
9. D
10. D

Page 5
1. B
2. C
3. A
4. B
5. D
6. B

Page 6
1. .63
2. .57
3. .53
4. .23
5. .67
6. .14

Page 7
1. C
2. B
3. C
4. A
5. D
6. D

Page 8
1. 0.2
2. 0.5
3. 0.1
4. 0.3
5. 0.8
6. 0.4
7. 0.6
8. 0.7
9. 0.9

Page 9
1. B
2. B
3. A
4. C
5. C
6. B
7. C
8. A
9. C
10. D

Page 10
1. A
2. D
3. D
4. D
5. B
6. A
7. C
8. A
9. B
10. B

Page 11
1. A
2. D
3. C
4. C
5. C
6. C
7. B
8. D
9. B
10. B

Page 12
1. D
2. C
3. C
4. D
5. D
6. B
7. B

Page 13
1. D
2. B
3. D
4. D

Page 14
1. B
2. D
3. B
4. B

Page 15
1. 8.5
2. 12.8
3. 11.4
4. 9.2
5. 15.2
6. 11.3
7. 14.0 or 14
8. 12.9
9. 11.6
10. 5.7
11. 12.7
12. 9.9
13. 10.5
14. 8.0 or 8

Page 16
1. 6.64
2. 7.46
3. 19.74
4. 33.4
5. 38.2
6. 82.8
7. 9.75
8. 11.6
9. 30.4

Page 17
1. 20.08
2. 15.50
3. 17.45
4. 15.25
5. 18.90
6. 157.61
7. 160.40
8. 15,614.87
9. 22.70
10. 15,584.63
11. 19.96
12. 16.17

Page 18
1. 32.83
2. 29.37
3. 41.01
4. 30.54
5. 42.83
6. 30.51
7. 34.12
8. 27.69
9. 34.87
10. 31.75
11. 56.53
12. 7.14

Page 19
1. B
2. B
3. D
4. A
5. A
6. C
7. C
8. B
9. C
10. D

Page 20
1. 3.26
2. 5.02
3. 3.01
4. 5.11
5. 8.68
6. 6.84
7. 2.85
8. 10.19
9. 1.57
10. 10.08

Page 21
1. 28.19
2. 4.60
3. 4.99
4. 13.90
5. 74.19
6. 13.42
7. 8.71
8. 11.0
9. 13.2
10. 12.4
11. 2.25
12. 107.8

Page 22
1. 68.5
2. 69.3
3. 91.6
4. 67.4
5. 75.76
6. 36.32
7. 83.3
8. 20.43
9. 57.3
10. 8.83
11. 49.35
12. 67.2

Page 23
1. 5, 6, 7, 8, 9
2. 0, 1, 2, 3, 4
3. 5, 6, 7, 8, 9
4. 0, 1, 2, 3, 4
5. 5, 6, 7, 8, 9
6. 0, 1, 2, 3, 4
7. 5, 6, 7, 8, 9
8. 0, 1, 2, 3, 4
9. 5, 6, 7, 8, 9
10. 0, 1, 2, 3, 4

Page 24
1. C
2. A
3. B
4. B
5. C
6. C
7. B
8. C

Page 25
1. 47
2. 71
3. 81
4. 66
5. 26
6. 46
7. 40
8. 34

Page 26
1. $1.05
2. $1.10
3. $1.20
4. $1.00

Page 27
1. C
2. C
3. D
4. A
5. A
6. D

Page 28
1. D
2. B
3. D
4. C
5. B

Page 29
1. C
2. C
3. B
4. D
5. A
6. Modeling Clay, Paint Set
7. Possible answer: two quarters; one dime; three pennies
8. Possible answer: one quarter; two dimes; one nickel; two pennies

Answer Key

9. Possible answer: one nickel; two pennies
10. Possible answer: three quarters; one dime; three pennies

Page 30
1. C
2. C
3. C
4. D
5. A
6. D
7. D
8. D
9. D
10. D

Page 31
1. B
2. A
3. C
4. A
5. C
6. A
7. B
8. D
9. A
10. A

Page 32
1. $9.57
2. $8.65
3. $17.55
4. $7.65
5. $9.01
6. $6.80
7. $16.42
8. $12.44
9. $10.30
10. $8.40

Page 33
1. $4.63
2. $4.80
3. $6.10
4. $6.21
5. $5.33
6. $5.95
7. $5.72
8. $5.29
9. $5.60
10. $5.95
11. $5.73
12. $5.24

Page 34
1. $65.17
2. $84.84
3. $89.19
4. $74.58
5. $51.76
6. $64.77
7. $69.38
8. $52.48
9. $98.99
10. $57.57
11. $73.49
12. $96.75

Page 35
1. B
2. A
3. C
4. D
5. C
6. D
7. B
8. C
9. D
10. A

Page 36
1. A
2. D
3. B
4. B
5. A
6. C
7. A
8. D
9. B
10. C
11. A
12. B
13. B
14. C

Page 37
1. D
2. A
3. C
4. A
5. D
6. D
7. A
8. B
9. B
10. A

Page 38
1. A
2. D
3. B
4. A
5. C
6. D
7. A
8. A
9. A
10. D

Page 39
1. $86.75
2. $206.88
3. $314.30
4. $374.75
5. $317.10
6. $253.68
7. $216.70
8. $374.31
9. $62.68
10. $187.20
11. $130.08
12. $95.41
13. $676.08
14. $34.35

Page 40
1. Ⓐ ● Ⓒ Ⓓ
2. ● Ⓑ Ⓒ Ⓓ
3. Ⓐ ● Ⓒ Ⓓ
4. Ⓐ Ⓑ Ⓒ ●
5. Ⓐ Ⓑ Ⓒ ●
6. Ⓐ ● Ⓒ Ⓓ
7. Ⓐ Ⓑ ● Ⓓ
8. Ⓐ Ⓑ Ⓒ ●
9. Ⓐ ● Ⓒ Ⓓ
10. ● Ⓑ Ⓒ Ⓓ
11. Ⓐ ● Ⓒ Ⓓ
12. ● Ⓑ Ⓒ Ⓓ
13. Ⓐ Ⓑ Ⓒ ●

Page 41
1. Ⓐ Ⓑ ● Ⓓ
2. ● Ⓑ Ⓒ Ⓓ
3. Ⓐ Ⓑ ● Ⓓ
4. ● Ⓑ Ⓒ Ⓓ
5. Ⓐ ● Ⓒ Ⓓ
6. Ⓐ Ⓑ Ⓒ ●
7. Ⓐ ● Ⓒ Ⓓ

Page 42
1. ● Ⓑ Ⓒ Ⓓ
2. Ⓐ Ⓑ Ⓒ ●
3. Ⓐ Ⓑ Ⓒ ●
4. ● Ⓑ Ⓒ Ⓓ
5. ● Ⓑ Ⓒ Ⓓ
6. ● Ⓑ Ⓒ Ⓓ
7. Ⓐ ● Ⓒ Ⓓ
8. ● Ⓑ Ⓒ Ⓓ
9. ● Ⓑ Ⓒ Ⓓ
10. Ⓐ Ⓑ ● Ⓓ
11. Ⓐ Ⓑ Ⓒ ●
12. Ⓐ Ⓑ Ⓒ ●
13. Ⓐ Ⓑ ● Ⓓ
14. Ⓐ Ⓑ Ⓒ ●
15. ● Ⓑ Ⓒ Ⓓ

Page 43
1. Ⓐ Ⓑ Ⓒ ● Ⓔ
2. Ⓐ Ⓑ Ⓒ ● Ⓔ
3. ● Ⓑ Ⓒ Ⓓ Ⓔ
4. Ⓐ Ⓑ Ⓒ ● Ⓔ
5. Ⓐ ● Ⓒ Ⓓ Ⓔ
6. Ⓐ Ⓑ Ⓒ Ⓓ ●
7. Ⓐ Ⓑ Ⓒ ● Ⓔ
8. Ⓐ ● Ⓒ Ⓓ Ⓔ
9. Ⓐ Ⓑ Ⓒ Ⓓ ●
10. ● Ⓑ Ⓒ Ⓓ Ⓔ

Page 44
1. Ⓐ Ⓑ ● Ⓓ Ⓔ
2. ● Ⓑ Ⓒ Ⓓ Ⓔ
3. Ⓐ ● Ⓒ Ⓓ Ⓔ
4. Ⓐ Ⓑ Ⓒ Ⓓ ●
5. Ⓐ Ⓑ Ⓒ ● Ⓔ
6. Ⓐ Ⓑ Ⓒ Ⓓ ●
7. Ⓐ Ⓑ Ⓒ Ⓓ ●
8. Ⓐ Ⓑ ● Ⓓ Ⓔ
9. ● Ⓑ Ⓒ Ⓓ Ⓔ
10. Ⓐ ● Ⓒ Ⓓ Ⓔ

Page 45
1. Ⓐ Ⓑ Ⓒ ●
2. Ⓐ ● Ⓒ Ⓓ
3. Ⓐ Ⓑ ● Ⓓ
4. Ⓐ Ⓑ Ⓒ ●
5. Ⓐ ● Ⓒ Ⓓ
6. Ⓐ Ⓑ Ⓒ ●
7. Ⓐ ● Ⓒ Ⓓ